Chemistry Research and Applications

Chemical Engineering Methods and Technology

Chromophores with Nonlinear Optical Properties and Their Applications
Maria Marinescu, PhD (Editor)
2023. ISBN: 979-8-88697-525-3 (Softcover)
2023. ISBN: 979-8-88697-726-4 (eBook)

Properties and Uses of Calcium Silicate
Amanda G. Carlton (Editor)
2022. ISBN: 979-8-88697-128-6 (Hardcover)
2022. ISBN: 979-8-88697-184-2 (eBook)

Chemical Vapor Deposition (CVD): Methods and Technologies
Levi Karlsson (Editor)
2021. ISBN: 978-1-53619-949-9 (Hardcover)
2021. ISBN: 978-1-53619-990-1 (eBook)

Fluidized Bed Reactors: Principles and Applications
Kristian Beike (Editor)
2020. ISBN: 978-1-53617-727-5 Softcover)
2020. ISBN: 978-1-53617-733-6 (eBook)

Microfiltration: Principles, Process and Applications
Youssef El Rayess, PhD (Editor)
2019. ISBN: 978-1-53616-320-9 (Hardcover)
2019. ISBN: 978-1-53616-347-6 (eBook)

More information about this series can be found at
https://novapublishers.com/product-category/series/chemical-engineering-methods-and-technology/

Calvin S. Willmon
Editor

Properties and Uses of Linolenic Acid

www.novapublishers.com

Copyright © 2023 by Nova Science Publishers, Inc.

All rights reserved. No part of this book may be reproduced, stored in a retrieval system or transmitted in any form or by any means: electronic, electrostatic, magnetic, tape, mechanical photocopying, recording or otherwise without the written permission of the Publisher.

We have partnered with Copyright Clearance Center to make it easy for you to obtain permissions to reuse content from this publication. Please visit copyright.com and search by Title, ISBN, or ISSN.

For further questions about using the service on copyright.com, please contact:

Copyright Clearance Center
Phone: +1-(978) 750-8400   Fax: +1-(978) 750-4470   E-mail: info@copyright.com

NOTICE TO THE READER

The Publisher has taken reasonable care in the preparation of this book but makes no expressed or implied warranty of any kind and assumes no responsibility for any errors or omissions. No liability is assumed for incidental or consequential damages in connection with or arising out of information contained in this book. The Publisher shall not be liable for any special, consequential, or exemplary damages resulting, in whole or in part, from the readers' use of, or reliance upon, this material. Any parts of this book based on government reports are so indicated and copyright is claimed for those parts to the extent applicable to compilations of such works.

Independent verification should be sought for any data, advice or recommendations contained in this book. In addition, no responsibility is assumed by the Publisher for any injury and/or damage to persons or property arising from any methods, products, instructions, ideas or otherwise contained in this publication.

This publication is designed to provide accurate and authoritative information with regards to the subject matter covered herein. It is sold with the clear understanding that the Publisher is not engaged in rendering legal or any other professional services. If legal or any other expert assistance is required, the services of a competent person should be sought. FROM A DECLARATION OF PARTICIPANTS JOINTLY ADOPTED BY A COMMITTEE OF THE AMERICAN BAR ASSOCIATION AND A COMMITTEE OF PUBLISHERS.

Library of Congress Cataloging-in-Publication Data

ISBN: 979-8-89113-201-6

Published by Nova Science Publishers, Inc. † New York

Contents

Preface		vii
Chapter 1	Improving and Preventing Hepatic Steatosis with α-Linolenic Acids	1
	Zahra Ahmadi, Navid Abedpoor and Fatemeh Hajibabaie	
Chapter 2	The Influence of Linolenic Acid on the Immune Response	21
	Elina Kaviani, Kamran Safavi, Fatemeh Hajibabaie and Navid Abedpoor	
Chapter 3	The Impacts of ALA-Based Natural Medications on Neurogenesis Statuses and Neurodegeneration Hallmarks	45
	Navid Abedpoor and Fatemeh Hajibabaie	
Chapter 4	Linolenic Acid Could Modulate the Pathomechanism Related to Colorectal Cancer	61
	Navid Abedpoor, Fatemeh Hajibabaie, Mohammad-Sajad Zare and Kamran Safavi	
Chapter 5	Natural Remedies that Offset Alpha-Linolenic Acid (ALA) Deficiency in Cardiovascular Hallmarks and Complications	77
	Fatemeh Hajibabaie, Navid Abedpoor and Elina Kaviani	
Index		95

Preface

This book contains five chapters that detail linolenic acid. Chapter One explains the effects α-linolenic acids (ALAs) have on improving and preventing hepatic steatosis. Chapter Two describes the effects of linolenic acid and its derivatives on the innate and adaptive immune response. Chapter Three overviews the numerous neurogenesis-attractive features of ALA and reviews the existing literature on the subject. Chapter Four explains how linolenic acid could modulate the pathomechanism related to colorectal cancer. Lastly, Chapter Five reviews the existing literature on ALA and gives a brief overview of the many cardiovascular attractive features of ALA.

Chapter 1

Improving and Preventing Hepatic Steatosis with α-Linolenic Acids

Zahra Ahmadi[1,2]
Navid Abedpoor[1,2],*
and Fatemeh Hajibabaie[1,3]
[1]Department of Physiology, Medicinal Plants Research Center,
Isfahan (Khorasgan) Branch, Islamic Azad University, Isfahan, Iran
[2]Department of Sports Physiology, Faculty of Sports Sciences,
Isfahan (Khorasgan) Branch, Islamic Azad University, Isfahan, Iran
[3]Department of Biology, Faculty of Basic Sciences, Shahrekord Branch,
Islamic Azad University, Shahrekord, Iran

Abstract

α-linolenic acids (ALA) is an essential fatty acid belonging to the omega-3 fatty acids group found in Flaxseed oil, sunflower, safflower, soybean, corn, and canola oils as well as nuts and seeds. Liver diseases, hepatic steatosis, type 2 diabetes, and alcoholic hepatic steatosis (AHS), might be triggered to the activation of pathways associated with inflammation and oxidative stress. While several treatments for these diseases exist, effective and appropriate approaches are not elucidated. ALA could help to manage and prevent diseases related to inflammation, oxidative stress, and alcoholic hepatic steatosis condition. Studies have shown that diets enriched with ALA supplementation effectively prevent hepatic steatosis. Moreover, ALA are associated with a reduction in insulin resistance, inflammation, and endoplasmic reticulum stress (ERS); and represent an alternative for improving liver function and obtaining

---

* Corresponding Author's Email: Abedpoor.navid@gmail.com, Abedpoor.navid@yahoo.com.

In: Properties and Uses of Linolenic Acid
Editor: Calvin S. Willmon
ISBN: 979-8-89113-201-6
© 2023 Nova Science Publishers, Inc.

metabolic benefits. ALA might be an effective approach for the management of liver diseases. ALA can decrease the fatty acid, lipid uptake. In this chapter, we will explain the effect of ALA on improving and preventing hepatic steatosis. Moreover, we will describe how consuming ALA can regulate inflammation and oxidative stress in the liver tissue and improve insulin resistance.

Keywords: liver, linolenic acid, endoplasmic reticulum stress, hepatic lipid, type 2 diabetes

1. Introduction

The prevalence of chronic liver disease is rapidly becoming a growing challenge for society (Méndez-Sánchez et al. 2022; Rahimi et al. 2021). There are about 71 million people around the world who suffer from liver disease. Improved comprehension of the pathophysiology of liver disease and advancements in diagnostic techniques, therapy, and preventive measures have significantly advanced clinical care for patients suffering from this condition (Williams 2006; Z. Younossi et al. 2018). As a result, approximately 1.5 billion people worldwide suffer from chronic liver disease (CLD) (Z.M. Younossi et al. 2023).

The varying degrees of metabolic activity in different age groups contribute to their susceptibility to drug toxicity (Lozupone et al. 2012). In addition, the harmful effects of drugs and environmental toxins on the liver can cause various clinical conditions (Zong and Friedman 2014).

Based on the evidence, nonalcoholic fatty liver disease (NAFLD) is the most common liver disease. NAFLD may lead to the development of cirrhosis, liver failure, and hepatocellular carcinoma (HCC) (Vieira Barbosa and Lai 2021). Moreover, alcoholic liver disease (ALD) is a significant health concern with notable physical, psychological, and financial consequences. This condition can exhibit various hepatic symptoms, such as alcoholic fatty liver disease, acute alcoholic hepatitis (AH), and alcoholic cirrhosis (Cohen 2016). Today, hepatitis C virus (HCV) infection remains one of the main causes of chronic liver disease. As the disease progresses, the risk of hepatic fibrosis, cirrhosis, and hepatocellular carcinoma (HCC) increases (Liver and Chair 2020).

On the other hand, the liver is one of the main organs that control metabolic homeostasis (Rahimi et al. 2021).

α-linolenic acids (ALA) are essential fatty acids (EFA), which means that, like vitamins, ALA must be obtained in the diet (Hajibabaie et al. 2022). The main role of ALA, is the precursor of eicosatetraenoic acid (EPA) and docosahexaenoic acid (DHA) (Valenzuela et al. 2012). ALA is a subset of N-3 fatty acid that plants can synthesize the basic n-6 and n-3 structures, but animals cannot (Calder 2012). There is a very high content of ALA in chloroplast membranes and certain seed oils (rapeseed, flaxseed, perilla seed, chia seed), beans (soybeans, navy beans), and nuts (walnuts) (Sinclair, Attar-Bashi, and Li 2002). Studies have shown that rats with ALA deficiency had the greatest changes in composition in the brain, heart, muscle, retina, and liver (Nock, Chouinard-Watkins, and Plourde 2017; Sinclair, Attar-Bashi, and Li 2002).

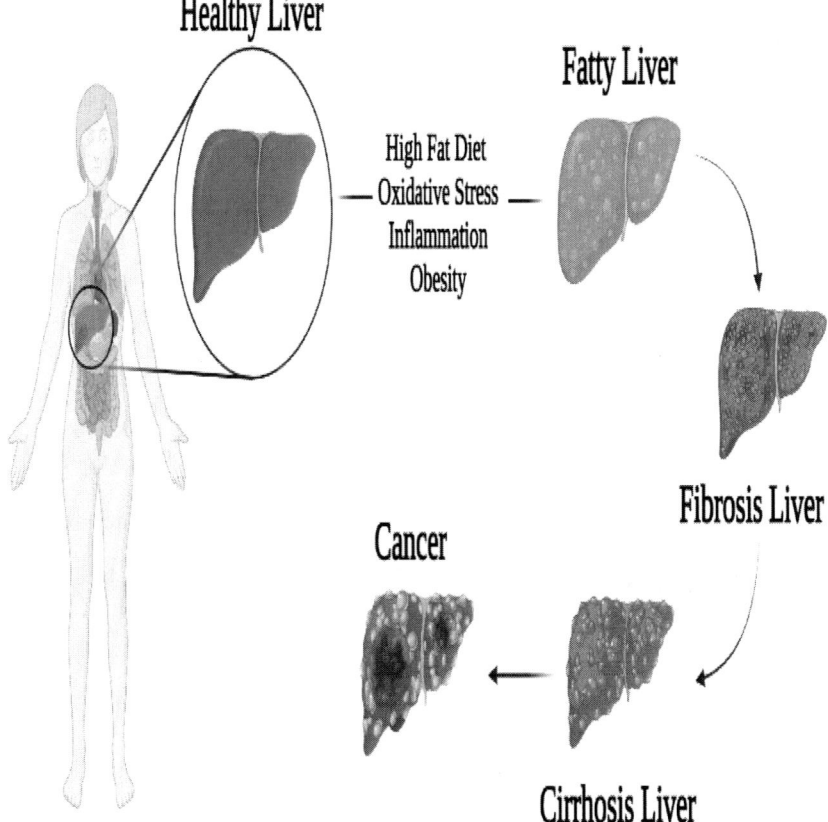

Figure 1. The progression of the liver disease via consumption of the high fat diet which lead to increase the inflammation and oxidative stress.

Fatty liver disease is a prevalent global ailment that arises from various factors such as dietary, environmental, genetic, and lifestyle-related factors (Rahimi et al. 2021). Several investigations have suggested that insulin resistance could coincide with an alteration in the fatty acid makeup in both bodily tissues and serum (Chou et al. 2023; Rahimi et al. 2021; Gonzalez et al. 2023). This alteration may be characterized by a shortage of n-3 polyunsaturated fatty acids (n-3 PUFA), long-chain fatty acids obtained from α-linolenic acid and found in fish oil (Calder 2006). These fatty acids are crucial for eicosanoid biosynthesis and interact with certain nuclear receptor proteins, leading to an impact on the transcription of regulatory genes (Capanni et al. 2006).

Storlien LH et al. proved that diets enriched with n-3 PUFA, especially ALA increase insulin sensitivity in rats (Storlien et al. 1987). Moreover, ALA affects several inflammatory pathways by modulating the expression of genes dependent on NF-κB (Rahimlou et al. 2019). In this chapter, we will describe the effect of ALA on liver diseases and diabetes.

2. The Association between Nonalcoholic Fatty Liver Disease and Alcoholic Hepatic Steatosis with ALA

Nonalcoholic fatty liver disease is notably correlated with a group of metabolic comorbidities comprising obesity, type 2 diabetes, high blood pressure, and increased levels of cholesterol. There are several proposed mechanisms to elucidate the pathogenesis and progression of nonalcoholic fatty liver disease (NAFLD) and increased oxidative stress (Abedpoor, Taghian, and Hajibabaie 2022). The growing occurrence of metabolic comorbidities augments the frequency of NAFLD and raises the possibility of patients developing more severe liver ailments (Z.M. Younossi et al. 2019). The presence of insulin resistance, oxidative stress, inflammation, and activation of the innate immune system all play a role in the development of nonalcoholic fatty liver disease (NAFLD). These factors promote the accumulation of fat and inflammation in the liver, leading to the development of steatosis and inflammation, which are key features of NAFLD (Figure 1) (Jung et al. 2012). At present, there is no pharmacological remedy that is considered effective for treating NAFLD (Rahimi et al. 2021). The only recommended approach for managing the condition is lifestyle changes, such as losing weight (Abedpoor, Taghian, and Hajibabaie 2022). Nonalcoholic fatty liver disease refers to a range of medical conditions that involve the nonalcoholic fatty liver, nonalcoholic steatohepatitis (NASH), cirrhosis, and associated complications (Z. Younossi et al. 2018).

Notably, changing the lifestyle of obese and overweight individuals might be difficult.

Therefore, complementary treatment and anti-inflammatory medications may be beneficial approaches to ameliorate the characteristics of NAFLD (Hajibabaie et al. 2022). ALA is a short-chain fatty acid with antioxidant and anti-inflammatory properties (Nabi et al. 2023). It is also an essential cofactor for multiple mitochondrial bioenergetics enzymes (Hajibabaie et al. 2022; Nabi et al. 2023). Moreover, studies have demonstrated that ALA can diminish levels of glucose, alleviate insulin resistance and oxidative stress, and regulate inflammatory signaling in different disorders (Lee et al. 2005). Amirkhizi and colleagues have shown that daily supplementation of ALA improves oxidative stress markers in patients with nonalcoholic fatty liver disease such as nonalcoholic fatty liver disease. Therefore, it may be regarded as a supplementary treatment option for preventing the progression of NAFLD (Amirkhizi et al. 2018).

Meng Wang has indicated that α-linolenic acid might suppress alcoholic hepatic steatosis by improving lipid homeostasis (Wang et al. 2016). Multiple mechanisms contribute to the pathogenesis of alcoholic fatty liver disease, including increased liver lipogenesis, altered mitochondrial fatty acid oxidation, and decreased export of very low-density lipoprotein (LDVL) (Altamirano and Bataller 2011). Therefore, it has been assumed that hepatic steatosis, which is characterized by an accumulation of lipids in the cytoplasm of hepatocytes, is the initial response to consuming alcohol and a reversible pathological condition (Altamirano and Bataller 2011). Moreover, Meng Wang established that α-linolenic acid improved lipid metabolisms in hepatic steatosis mice. In addition, adipose triglyceride lipase (ATGL) and phosphor-hormone-sensitive lipase (HSL) have a crucial role in adipose lipolysis (Roberts et al. 2002). Therefore, α-linolenic acid can increase the ATGL and HSL expression levels. Furthermore, CD36, as vital fatty acids transporter, could regulate the fatty acids uptake in the liver and adipose tissue (Wang et al. 2016). Lipoprotein lipase (Lpl), very low-density lipoprotein receptor (Vldl-r), and Fatbp1 are regulated by the consumption of α-linolenic acid (Wang et al. 2016). Interestingly, consumption of the α-linolenic acid regulated the expression level of the Pparγ, adenosine monophosphate-activated protein kinase (AMPK), acetyl-CoA carboxylase (ACC), and adiponectin receptor (ADIPOR2) (Wang et al. 2016). In another study, Natalia Bonissi Gonc indicated that α-linolenic acid supplementation decreased insulin resistance and ameliorated glucose concentration (Gonçalves et al. 2018). Moreover, Natalia Bonissi

Gonc and co-workers have indicated that α-linolenic acid supplementation could decline the serum concentration of the interleukin one beta (IL-1β), interleukin 6 (IL-6), monocyte chemoattractant protein-1 (MCP-1) (Gonçalves et al. 2018). Furthermore, based on these data, the expression level of the essential chaperones, including heat shock protein 70 (HSP70), significantly increased the binding immunoglobulin protein (BIP) (Gonçalves et al. 2018). In addition, the expression level of the X-box binding protein 1 (XBP1) and C/EBP-homologous protein (CHOP) remarkably decreased in hepatic tissues (Figure 2) (Gonçalves et al. 2018).

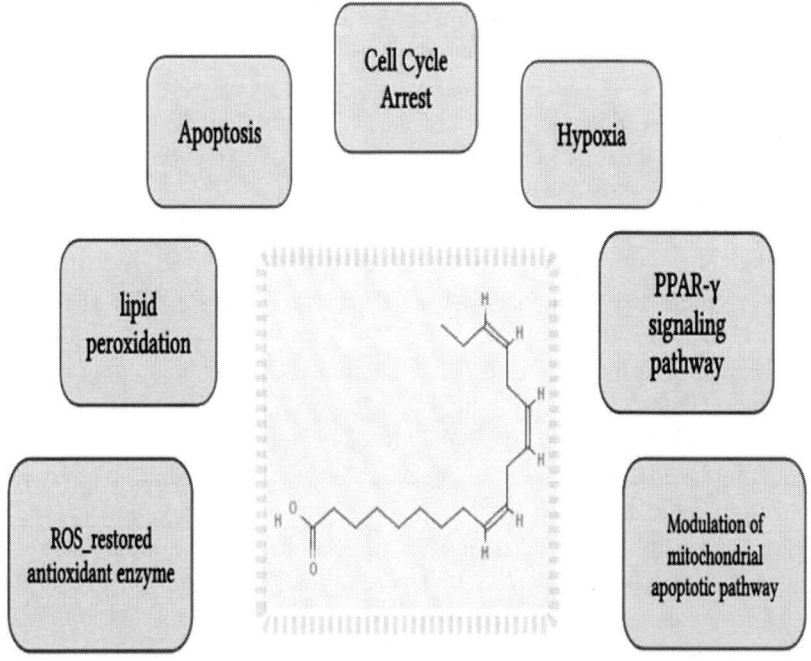

Figure 2. α-linolenic acids (ALA) can modulate the apoptosis, hypoxia, Ppary signaling pathway, oxidative stress, lipid peroxidation, and mitochondrial apoptotic pathway.

Recent evidence demonstrates that endoplasmic reticulum (ER) stress is a crucial factor in the development of NAFLD (Ashraf and Sheikh 2015). In addition, the ER is the primary organ responsible for protein folding, lipid biogenesis, and calcium homeostasis (Kaneko et al. 2017).

Even though the unfolded protein response (UPR) can be activated by ER stress as a means of restoring ER homeostasis and cell survival, prolonged ER stress is known to lead to negative outcomes such as abnormal fat storage, inflammation, oxidative stress, apoptosis, and disruptions in autophagy (Li et al. 2015).

All the processes mentioned above can trigger the onset of NAFLD (Lebeaupin et al. 2018). In addition, recent research revealed that NAFLD is caused by ER stress-mediated mitochondrial dysfunction (Gao et al. 2018). Moreover, ER stress activates increasing reactive oxygen species (ROS) and ER oxidoreductase 1 (ERO1). The ROS-induced activation of the ER protein inositol-1,4,5-trisphosphate receptors, combined with the inactivation of the sarcoplasmic reticulum Ca2+-ATPase, leads to a rise in cytosolic Ca2+, an increase in mitochondrial Ca2+ uptake and finally mitochondrial malfunction (Dandekar, Mendez, and Zhang 2015). Mitochondria is crucial in FA degradation as the primary site for fatty acid (FA) oxidation. Therefore, a decrease in the oxidation of fatty acids in mitochondria can play a role in the fatty accumulation in the liver and the occurrence of NAFLD (Dandekar, Mendez, and Zhang 2015; Peng et al. 2018). ALA, an n-3 PUFA found in plants, has a powerful influence on reducing TG, promoting FA catabolism, and blocking inflammation (Miotto et al. 2017; Arendt et al. 2015). Accordingly, plant sterol has recently been esterified by ALA to produce a final product, plant ester of -linolenic acid (PS-ALA), which simultaneously reduces TC and TG, thereby enhancing ER homeostasis and protecting against NAFLD (Arendt et al. 2015; Han et al. 2019). Unfortunately, no pharmacological treatment for NAFLD has been identified, and thus a healthy diet and regular physical activity are recommended (Leoni et al. 2018). However, recent research has investigated various natural materials or phytochemicals with the potential to regulate lipids, act as antioxidants, and reduce inflammation (Leoni et al. 2018; Yu et al. 2018). Based on the data mining, the intervention ALA can potentially reduce serum levels of inflammatory cytokines (IL-1β, IL-6, TNF-α, and MCP-1) and attenuate hepatic steatosis by regulating ER stress (Yu et al. 2018).

Table 1. The genes with high degree, betweenness centrality, closeness centrality, and clustering coefficient involved in anti-liver disease effects

Symbole genes	Degree	Betweenness Centrality	Closeness Centrality	Clustering Coefficient
ABHD5	6	1.30E-05	0.38	0.93333333
LIPE	16	0.00667501	0.45783133	0.55833333
PPARGC1A	35	0.06274667	0.53900709	0.38319328
PPARG	35	0.06940326	0.55072464	0.35630252
PLIN1	12	0.00117338	0.39378238	0.71212121
PNPLA2	10	0.00232855	0.44186047	0.73333333

Symbole genes	Degree	Betweenness Centrality	Closeness Centrality	Clustering Coefficient
PPARA	29	0.02573707	0.51351351	0.3817734
AIM2	4	1.17E-04	0.32478632	0.83333333
PYCARD	3	0	0.32340426	1
IL18	12	0.06395412	0.46625767	0.37878788
NLRP3	10	0.03139885	0.44970414	0.4
NEK7	3	0	0.32340426	1
ARNTL	12	3.82E-04	0.45508982	0.84848485
SIRT1	34	0.21593373	0.6440678	0.36541889
CREBBP	22	0.00768397	0.49350649	0.64502165
EP300	24	0.01056153	0.5	0.58333333
NCOR1	19	0.00295273	0.48407643	0.71345029
TP53	37	0.23461362	0.66086957	0.35435435
CLOCK	10	1.12E-04	0.44444444	0.91111111
EZH2	15	0.00122648	0.4691358	0.78095238
HDAC1	20	0.03962294	0.56296296	0.58421053
ESRRA	13	0.00112686	0.46060606	0.73076923
ATF6	11	0.00200421	0.44705882	0.63636364
XBP1	13	0.00239049	0.45783133	0.69230769
YY1	17	0.04028349	0.55072464	0.57352941
HSP90B1	7	0	0.41758242	1
EIF2AK3	12	0.00286859	0.45238095	0.60606061
HSPA5	13	0.00327437	0.46341463	0.64102564
ERN1	10	0.00411882	0.44970414	0.64444444
DDIT3	12	0.00160574	0.44705882	0.66666667
BMS1	25	0	0.41304348	1
FBL	27	0.11953373	0.54676259	0.86039886
UTP18	25	0	0.41304348	1
RRP9	26	0.07597543	0.52777778	0.92615385
MPHOSPH10	25	0	0.41304348	1
UTP3	25	0	0.41304348	1
IMP4	25	0	0.41304348	1
UTP20	25	0	0.41304348	1
UTP6	25	0	0.41304348	1
NOP58	25	0	0.41304348	1
PWP2	25	0	0.41304348	1
UTP15	25	0	0.41304348	1
DCAF13	25	0	0.41304348	1
NOL6	25	0	0.41304348	1
WDR3	25	0	0.41304348	1
WDR75	25	0	0.41304348	1

Serum concentrates and protein expressions of the Nlrp3 inflammasome, Il-18, and Il-1β were elevated by HFD, along with phosphorylation activation of Ire1a and increased mRNA and protein expression of Xbp1s (Han et al. 2019). Chronic ER stress has been shown to initiate oxidative stress by increasing ER $Ca2+$ release and oxidoreductase-1 activity, both leading to mitochondrial injury (Chen et al. 2017). FA decomposition is primarily catalyzed by -oxidation occurring primarily in mitochondria. Mitochondrial dysfunction coincides with suboptimal or incomplete fat oxidation, accumulating TG, which can lead to hepatic steatosis (Simões et al. 2018). In human and rodent models of NAFLD, mitochondrial dysfunction is a fundamental feature of the transition from SFL to NASH (Sunny, Bril, and Cusi 2017).

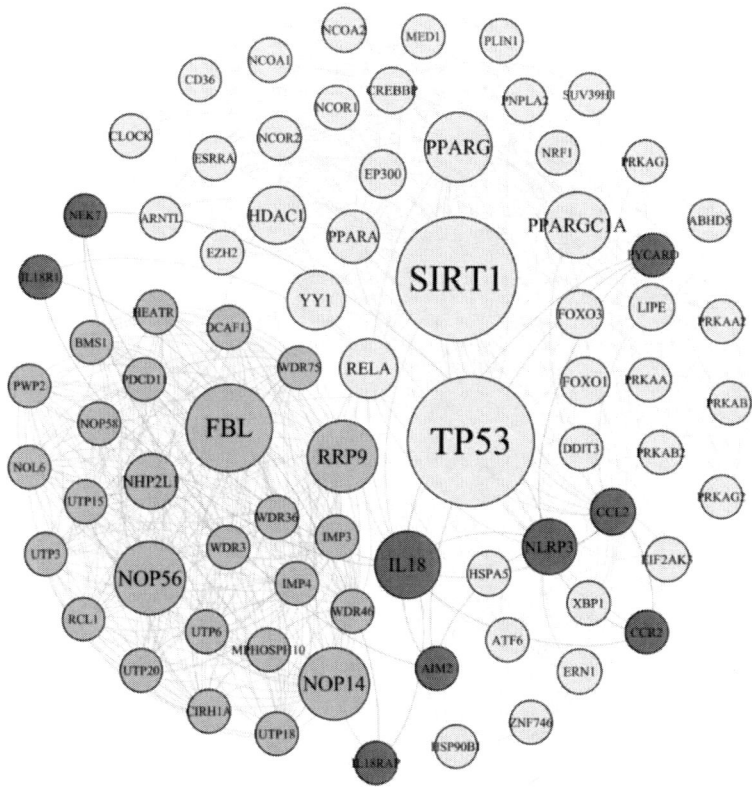

Figure 3. The protein- protein interaction involved in the multiple biological mechanisms underlying anti-liver disease effects.

Therefore, improving mitochondrial activity could be a potential treatment for NAFLD. It has been verified that increased proteins responsible for mitochondrial biogenesis can increase oxidative metabolic capacity (Sunny, Bril, and Cusi 2017). The upregulated transcription of genes related to oxidative phosphorylation by PGC-1 can activate Nrf1, a transcription factor responsible for regulating nuclear genes involved in coding mitochondrial respiratory chain proteins. The transcription and replication of mtDNA are maintained and regulated by the cooperative action of PGC-1 and Nrf1 (Han et al. 2019; Kim et al. 2015).

We indicated the protein-protein interactions involved in the multiple biological mechanisms underlying the anti-liver disease effects (Table 1, Figure 3)

3. Diabetes and ALA

Epidemiology evidence has indicated that there has been a significant increase in the prevalence of diabetes mellitus (DM) (Glovaci, Fan, and Wong 2019). Type 2 diabetes (T2D) is defined by pancreatic β-cell dysfunction due to insulin production and insulin resistance in target organs (Hajibabaie et al. 2022; Kim et al. 2015; Abedpoor, Taghian, and Hajibabaie 2022).

The evidence has demonstrated that ALA can regulate blood sugar. Based on these data, ALA consumption reduced glucose concentration and enhanced insulin sensitivity which led to decreased risk of T2D (Seah et al. 2019; Jiang et al. 2021)

ALA has been shown in multiple animal and laboratory experiments to have the capability to regulate glucose balance by impacting insulin sensitivity through possible roles in controlling gene expression, managing fat metabolism, and influencing the creation of fat cells (Rodríguez-Correa et al. 2020). Furthermore, activation of PPARγ in individuals with T2D leads to notable enhancements in insulin sensitivity, which in turn may result in improvements in various glycemic control measures (Hansson et al. 2020).

Due to the fact that PPARγ plays a crucial role in regulating the metabolism of adipose tissue, ALA can amplify the expression level of PPARγ (Roglans et al. 2021). In addition, an ALA-rich diet has been shown to reduce oxidative stress in humans and rats involved with hypercholesterolemia (Xie et al. 2011).

Conclusion

α-linolenic acids could control and manage liver disease via modulating the apoptosis, hypoxia, Pparγ signaling pathway, oxidative stress, lipid peroxidation, and mitochondrial apoptotic pathway. In addition, α-linolenic acids could ameliorate the expression level of the Nlrp3, Il-18, PGC-1, Nrf1, and Il-1β.

References

Abedpoor, Navid, Farzaneh Taghian, and Fatemeh Hajibabaie. (2022). "Physical activity ameliorates the function of organs via adipose tissue in metabolic diseases." *Acta histochemical*, 124 (2), 151844.

Altamirano, Jose, and Ramon Bataller. (2011). "Alcoholic liver disease: pathogenesis and new targets for therapy." *Nature reviews Gastroenterology & hepatology*, 8 (9), 491-501.

Amirkhizi, Farshad, Soudabeh Hamedi-Shahraki, Sonya Hosseinpour-Arjmand, Elnaz Vaghef-Mehrabany, and Mehrangiz Ebrahimi-Mameghani. (2018). "Effects of alpha-lipoic acid supplementation on oxidative stress status in patients with non-alcoholic fatty liver disease: a randomized, double-blind, placebo-controlled clinical trial." *Iranian Red Crescent Medical Journal*, 20 (9), 11.

Arendt, Bianca, M., Elena M Comelli, David WL Ma, Wendy Lou, Anastasia Teterina, TaeHyung Kim, Scott K Fung, David KH Wong, Ian McGilvray, and Sandra E Fischer. (2015). "Altered hepatic gene expression in nonalcoholic fatty liver disease is associated with lower hepatic n-3 and n-6 polyunsaturated fatty acids." *Hepatology*, 61 (5), 1565-1578.

Ashraf, N. U., and Sheikh, T. A. (2015). "Endoplasmic reticulum stress and oxidative stress in the pathogenesis of non-alcoholic fatty liver disease." *Free radical research*, 49 (12), 1405-1418.

Calder, Philip, C. (2006). "n− 3 Polyunsaturated fatty acids, inflammation, and inflammatory diseases." *The American journal of clinical nutrition*, 83 (6), 1505S-1519S.

---. (2012). "Mechanisms of action of (n-3) fatty acids." *The Journal of nutrition*, 142 (3), 592S-599S.

Capanni, M., Calella, F., Biagini, M. R., Genise, S., Raimondi, L., Bedogni, G., Svegliati-Baroni, G., Sofi, F., Milani, S., and Abbate, R. (2006). "Prolonged n-3 polyunsaturated fatty acid supplementation ameliorates hepatic steatosis in patients with non-alcoholic fatty liver disease: a pilot study." *Alimentary pharmacology & therapeutics*, 23 (8), 1143-1151.

Chen, Qun, Jeremy Thompson, Ying Hu, Anindita Das, and Edward J Lesnefsky. (2017). "Metformin attenuates ER stress–induced mitochondrial dysfunction." *Translational Research*, 190, 40-50.

Chou, Tzu-Jung, Chia-Wen Lu, Li-Yu Lin, Yi-Ju Hsu, Chi-Chang Huang, and Kuo-Chin Huang. (2023). "Proteomic Analysis of Skeletal Muscle and White Adipose Tissue after Aerobic Exercise Training in High Fat Diet Induced Obese Mice." *International Journal of Molecular Sciences*, 24 (6), 5743.

Cohen, Stanley Martin. (2016). "Alcoholic liver disease." *Clinics in liver disease*, 20 (3), xiii-xiv.

Dandekar, Aditya, Roberto Mendez, and Kezhong Zhang. (2015). "Cross talk between ER stress, oxidative stress, and inflammation in health and disease." *Stress Responses: Methods and Protocols*, 205-214.

Gao, Xiaobo, Shun Guo, Song Zhang, An Liu, Lei Shi, and Yan Zhang. 2018. "Matrine attenuates endoplasmic reticulum stress and mitochondrion dysfunction in nonalcoholic fatty liver disease by regulating SERCA pathway." *Journal of Translational Medicine*, 16 (1), 1-16.

Glovaci, Diana, Wenjun Fan, and Nathan D Wong. (2019). "Epidemiology of diabetes mellitus and cardiovascular disease." *Current cardiology reports*, 21, 1-8.

Gonçalves, Natália Bonissi, Rafael Ferraz Bannitz, Bruna Ramos Silva, Danielle Duran Becari, Carolina Poloni, Patrícia Moreira Gomes, Milton Cesar Foss, and Maria Cristina Foss-Freitas. (2018). "α-Linolenic acid prevents hepatic steatosis and improves glucose tolerance in mice fed a high-fat diet." *Clinics*, 73.

Gonzalez, Patrick, Alexandre Dos Santos, Marion Darnaud, Nicolas Moniaux, Delphine Rapoud, Claire Lacoste, Tung-Son Nguyen, Valentine S Moullé, Alice Deshayes, and Gilles Amouyal. (2023). "Antimicrobial protein REG3A regulates glucose homeostasis and insulin resistance in obese diabetic mice." *Communications Biology*, 6 (1), 269.

Hajibabaie, Fatemeh, Navid Abedpoor, Kamran Safavi, and Farzaneh Taghian. (2022). "Natural remedies medicine derived from flaxseed (secoisolariciresinol diglucoside, lignans, and α-linolenic acid) improve network targeting efficiency of diabetic heart conditions based on computational chemistry techniques and pharmacophore modeling." *Journal of Food Biochemistry*, e14480.

Han, Hao, Yan Guo, Xiaoyu Li, Dongxing Shi, Tingli Xue, Linqi Wang, Yanyan Li, and Mingming Zheng. (2019). "Plant sterol ester of α-linolenic acid attenuates nonalcoholic fatty liver disease by rescuing the adaption to endoplasmic reticulum stress and enhancing mitochondrial biogenesis." *Oxidative medicine and cellular longevity*, 2019.

Hansson, Björn, Sara Schumacher, Claes Fryklund, Björn Morén, Maria Björkqvist, Karl Swärd, and Karin G Stenkula. (2020). "A hypothesis for insulin resistance in primary human adipocytes involving MRTF-A and suppression of PPARγ." *Biochemical and Biophysical Research Communications*, 533 (1), 64-69.

Jiang, Hong, Lina Wang, Duolao Wang, Ni Yan, Chao Li, Min Wu, Jing Lin, Wanru Jia, Xi Liu, and Jiaxin Lv. (2021). "Omega-3 Polyunsaturated Fatty Acid Biomarkers and Risk of Type 2 Diabetes, Cardiovascular Disease, Cancer, and Mortality: A Systematic Review and Meta-analysis."

Jung, Tae Sik, Soo Kyoung Kim, Hyun Joo Shin, Byeong Tak Jeon, Jong Ryeal Hahm, and Gu Seob Roh. (2012). "α-lipoic acid prevents non-alcoholic fatty liver disease in OLETF rats." *Liver International*, 32 (10), 1565-1573.

Kaneko, Masayuki, Kazunori Imaizumi, Atsushi Saito, Soshi Kanemoto, Rie Asada, Koji Matsuhisa, and Yosuke Ohtake. (2017). "ER stress and disease: toward prevention and treatment." *Biological and Pharmaceutical Bulletin*, 40 (9), 1337-1343.

Kim, Chu-Sook, Yoonhee Kwon, Suck-Young Choe, Sun-Myung Hong, Hoon Yoo, Tsuyoshi Goto, Teruo Kawada, Hye-Seon Choi, Yeonsoo Joe, and Hun Taeg Chung. (2015). "Quercetin reduces obesity-induced hepatosteatosis by enhancing mitochondrial oxidative metabolism via heme oxygenase-1." *Nutrition & metabolism*, 12 (1), 1-9.

Lebeaupin, Cynthia, Deborah Vallée, Younis Hazari, Claudio Hetz, Eric Chevet, and Béatrice Bailly-Maitre. (2018). "Endoplasmic reticulum stress signalling and the pathogenesis of non-alcoholic fatty liver disease." *Journal of hepatology*, 69 (4), 927-947.

Lee, Woo Je, Kee-Ho Song, Eun Hee Koh, Jong Chul Won, Hyoun Sik Kim, Hye-Sun Park, Min-Seon Kim, Seung-Whan Kim, Ki-Up Lee, and Joong-Yeol Park. (2005). "α-Lipoic acid increases insulin sensitivity by activating AMPK in skeletal muscle." *Biochemical and biophysical research communications*, 332 (3), 885-891.

Leoni, Simona, Francesco Tovoli, Lucia Napoli, Ilaria Serio, Silvia Ferri, and Luigi Bolondi. (2018). "Current guidelines for the management of non-alcoholic fatty liver disease: A systematic review with comparative analysis." *World journal of gastroenterology*, 24 (30), 3361.

Li, Xiaohui, Yarui Wang, Huan Wang, Cheng Huang, Yan Huang, and Jun Li. (2015). "Endoplasmic reticulum stress is the crossroads of autophagy, inflammation, and apoptosis signaling pathways and participates in liver fibrosis." *Inflammation Research*, 64, 1-7.

Liver, European Association for the Study of the, and Clinical Practice Guidelines Panel: Chair. (2020). "EASL recommendations on treatment of hepatitis C: final update of the series." *J. Hepatol.*, 73, 1170-1218.

Lozupone, Catherine, A., Jesse I Stombaugh, Jeffrey I Gordon, Janet K Jansson, and Rob Knight. (2012). "Diversity, stability and resilience of the human gut microbiota." *Nature*, 489 (7415), 220-230.

Méndez-Sánchez, Nahum, Elisabetta Bugianesi, Robert G Gish, Frank Lammert, Herbert Tilg, Mindie H Nguyen, Shiv K Sarin, Núria Fabrellas, Shira Zelber-Sagi, and Jian-Gao Fan. (2022). "Global multi-stakeholder endorsement of the MAFLD definition." *The lancet Gastroenterology & hepatology*, 7 (5), 388-390.

Miotto, Paula, M., Meaghan Horbatuk, Ross Proudfoot, Sarthak Matravadia, Marica Bakovic, Adrian Chabowski, and Graham P Holloway. (2017). "α-Linolenic acid supplementation and exercise training reveal independent and additive responses on hepatic lipid accumulation in obese rats." *American Journal of Physiology-Endocrinology and Metabolism*.

Nabi, Oumarou, Nathanaël Lapidus, Jérôme Boursier, Victor de Ledinghen, Jean-Michel Petit, Sofiane Kab, Adeline Renuy, Marie Zins, Karine Lacombe, and Lawrence Serfaty. (2023). "Lean individuals with NAFLD have more severe liver disease and poorer clinical outcomes (NASH-CO Study)." *Hepatology*, 10.1097.

Nock, Tanya Gwendolyn, Raphaël Chouinard-Watkins, and Mélanie Plourde. (2017). "Carriers of an apolipoprotein E epsilon 4 allele are more vulnerable to a dietary deficiency in omega-3 fatty acids and cognitive decline." *Biochimica et Biophysica Acta (BBA)-Molecular and Cell Biology of Lipids*, 1862 (10), 1068-1078.

Peng, Kang-Yu, Matthew J Watt, Sander Rensen, Jan Willem Greve, Kevin Huynh, Kaushala S Jayawardana, Peter J Meikle, and Ruth CR Meex. (2018). "Mitochondrial dysfunction-related lipid changes occur in nonalcoholic fatty liver disease progression." *Journal of lipid research*, 59 (10), 1977-1986.

Rahimi, Golbarg, Salime Heydari, Bahareh Rahimi, Navid Abedpoor, Iman Niktab, Zahra Safaeinejad, Maryam Peymani, Farzad Seyed Forootan, Zahra Derakhshan, and Mohammad Hossein Nasr Esfahani. (2021). "A combination of herbal compound (SPTC) along with exercise or metformin more efficiently alleviated diabetic complications through down-regulation of stress oxidative pathway upon activating Nrf2-Keap1 axis in AGE rich diet-induced type 2 diabetic mice." *Nutrition & metabolism*, 18, 1-14.

Rahimlou, Mehran, Maryam Asadi, Nasrin Banaei Jahromi, and Anahita Mansoori. (2019). "Alpha-lipoic acid (ALA) supplementation effect on glycemic and inflammatory biomarkers: A Systematic Review and meta-analysis." *Clinical nutrition ESPEN*, 32, 16-28.

Roberts, Christian, K., James Barnard, R., Kai Hui Liang, and Nosratola D Vaziri. (2002). "Effect of diet on adipose tissue and skeletal muscle VLDL receptor and LPL: implications for obesity and hyperlipidemia." *Atherosclerosis*, 161 (1), 133-141.

Rodríguez-Correa, Eduardo, Imelda González-Pérez, Pedro Isauro Clavel-Pérez, Yolanda Contreras-Vargas, and Karla Carvajal. (2020). "Biochemical and nutritional overview of diet-induced metabolic syndrome models in rats: what is the best choice?" *Nutrition & diabetes*, 10 (1), 24.

Roglans, Nuria, Miguel Baena, Gemma Sangüesa, Ana Magdalena Velázquez, Christian Griñán-Ferré, Mercè Pallàs, Rosa María Sánchez, Marta Alegret, and Juan Carlos Laguna. (2021). "Chronic liquid fructose supplementation does not cause liver tumorigenesis but elicits clear sex differences in the metabolic response in Sprague–Dawley rats." *Food & Nutrition Research*, 65.

Seah, Jowy Y. H., Choon Nam Ong, Woon-Puay Koh, Jian-Min Yuan, and Rob M van Dam. (2019). "A dietary pattern derived from reduced rank regression and fatty acid biomarkers is associated with lower risk of type 2 diabetes and coronary artery disease in Chinese adults." *The Journal of nutrition*, 149 (11), 2001-2010.

Simões, Inês, C. M., Adriana Fontes, Paolo Pinton, Hans Zischka, and Mariusz R Wieckowski. (2018). "Mitochondria in non-alcoholic fatty liver disease." *The international journal of biochemistry & cell biology*, 95, 93-99.

Sinclair, Andrew, J., Nadia M Attar-Bashi, and Duo Li. (2002). "What is the role of α-linolenic acid for mammals?" *Lipids*, 37 (12), 1113-1123.

Storlien, Leonard, H., Edward W Kraegen, Donald J Chisholm, Glenn L Ford, David G Bruce, and Wendy S Pascoe. (1987). "Fish oil prevents insulin resistance induced by high-fat feeding in rats." *Science*, 237 (4817), 885-888.

Sunny, Nishanth, E., Fernando Bril, and Kenneth Cusi. (2017). "Mitochondrial adaptation in nonalcoholic fatty liver disease: novel mechanisms and treatment strategies." *Trends in Endocrinology & Metabolism*, 28 (4), 250-260.

Valenzuela, R., Gormáz, J. G., Masson, L., Vizcarra, M, and Cornejo, P. (2012). "Evaluation of the hepatic bioconversion of α-linolenic acid (ALA) to eicosapentaenoic acid (EPA) and docosahexaenoic acid (DHA) in rats fed with oils from chia (Salvia hispánica) or rosa mosqueta (Rosa rubiginosa)." *Grasasyaceites*, 63, 1.

Vieira Barbosa, Joana, and Michelle Lai. (2021). "Nonalcoholic fatty liver disease screening in type 2 diabetes mellitus patients in the primary care setting." *Hepatology Communications*, 5 (2), 158-167.

Wang, Meng, Xiao-Jing Zhang, Kun Feng, Chengwei He, Peng Li, Yuan-Jia Hu, Huanxing Su, and Jian-Bo Wan. (2016). "Dietary α-linolenic acid-rich flaxseed oil prevents against alcoholic hepatic steatosis via ameliorating lipid homeostasis at adipose tissue-liver axis in mice." *Scientific Reports*, 6 (1), 1-11.

Williams, Roger. (2006). "Global challenges in liver disease." *Hepatology*, 44 (3), 521-526.

Xie, Nianlin, Wei Zhang, Jia Li, Hongliang Liang, Huasong Zhou, Weixun Duan, Xuezeng Xu, Shiqiang Yu, Haifeng Zhang, and Dinghua Yi. (2011). "α-Linolenic acid intake attenuates myocardial ischemia/reperfusion injury through anti-inflammatory and anti-oxidative stress effects in diabetic but not normal rats." *Archives of Medical Research*, 42 (3), 171-181.

Younossi, Zobair, Quentin M Anstee, Milena Marietti, Timothy Hardy, Linda Henry, Mohammed Eslam, Jacob George, and Elisabetta Bugianesi. (2018). "Global burden of NAFLD and NASH: trends, predictions, risk factors and prevention." *Nature reviews Gastroenterology & hepatology*, 15 (1), 11-20.

Younossi, Zobair, M., Pegah Golabi, James M Paik, Austin Henry, Catherine Van Dongen, and Linda Henry. (2023). "The global epidemiology of nonalcoholic fatty liver disease (NAFLD) and nonalcoholic steatohepatitis (NASH): A systematic review." *Hepatology*, 77 (4), 1335-1347.

Younossi, Zobair, M., Giulio Marchesini, Helena Pinto-Cortez, and Salvatore Petta. (2019). "Epidemiology of nonalcoholic fatty liver disease and nonalcoholic steatohepatitis: implications for liver transplantation." *Transplantation*, 103 (1), 22-27.

Yu, Xiao, Qianchun Deng, Yuhan Tang, Lin Xiao, Liegang Liu, Ping Yao, Hu Tang, and Xuyan Dong. (2018). "Flaxseed oil attenuates hepatic steatosis and insulin resistance in mice by rescuing the adaption to ER stress." *Journal of agricultural and food chemistry*, 66 (41), 10729-10740.

Zong, Yiwei, and Joshua R Friedman. (2014). "Liver development." *Liver Disease in Children*, 1-813.

Biographical Sketch

Name: Navid Abedpoor
Affiliation: Department of Sports Physiology, Faculty of Sports Sciences, Isfahan (Khorasgan) Branch, Islamic Azad University, Isfahan, Iran.

Department of Physiology, Medicinal Plants Research Center, Isfahan (Khorasgan) Branch, Islamic Azad University, Isfahan, Iran.

Education: Sport sciences

Business Address: Isfahan (Khorasgan) Branch, Islamic Azad University, Isfahan, Iran.

Research and Professional Experience: Inflammation, Oxidative Stress, Non coding RNAs, Cancer, Lifestyles

Professional Appointments: Researcher

Honors:
1. **Won the Young Scientist** Award in "International Scientist Awards on Engineering, Science, and Medicine." 2020.
2. **Best Poster** in Royan International twin congress, Reproductive Biomedicine & Stem Cell. Protective approaches of Fraxinus excelsior compounds on the Implantation based infertility via bioinformatics and chemoinformatic analysis.

Publications from the Last 3 Years:

[1] Fatemeh Azizian-Farsani, Navid Abedpoor, Mohammad Hasan Sheikhha, Ali Osmay Gure, Mohammad Hossein Nasr Esfahani, and Kamran Ghaedi. (2020). Receptor for advanced glycation end products acts as a fuel to colorectal cancer development. *Frontiers in Oncology*, IF: 6.5.

[2] Fatemeh Azizian-Farsani, Marcin Osuchowski, Navid Abedpoor, Farzad Seyed Forootan, Maryam Derakhshan, Mohammad Hossein Nasr-Esfahani, Mohammad Hasan Sheikhha, and Kamran Ghaedi. (2020). Anti-inflammatory and -apoptotic effects of an herbal extract on DSS-induced colitis in mice fed with high AGEs-fat diet. Scientific Reports. *Nutrition & Metabolism*, IF: 4.5.

[3] Golbarg Rahimi; Salime Heydari; Bahare Rahimi; Navid Abedpoor; Iman Nicktab; Zahra Safaeinejad; Maryam Peymani; Farzad Seyed Forootan; Zahra Derakhshan; Mohammad Hossein Nasr Esfahani, and Kamran Ghaedi. (2020). A combination of herbal compound (SPTC) along with exercise or metformin more efficiently alleviated diabetic complications through down-regulation of stress oxidative pathway upon activating Nrf2-Keap1 axis in AGEs rich diet-induced type 2 diabetic mice. *Nutrition and metabolism*, IF:4.5.

[4] Fahimeh Akbarian, Mohsen Rahmani, Marziyeh Tavalaee, Navid Abedpoor, Mozhdeh Taki, Kamran Ghaedi, and Mohammad Hossein Nasr-Esfahani. (2021). Effect of different high-fat and AGEs diets in obesity and diabetes-prone C57BL/6 mice on sperm function. *International Journal of Fertility and Sterility*, IF:2.7.

[5] Navid Abedpoor, Farzaneh Taghian, and Fatemeh Hajibabaie. (2022). Physical activity ameliorates the function of organs via adipose tissue in metabolic diseases. *Acta histochemical*, IF: 2.7.

[6] Fatemeh Hajibabaie, Navid Abedpoor, Nazanin Asareh, Mohammad Amin Tabatabaiefar, Ali Zarrabi and Laleh Shariati. (2022). A cocktail of microRNAs as an advance diagnostic signature in stomach-colorectal cancers hallmarks incidence: a systematic review. Personal Medicine.

[7] Navid Abedpoor, Farzaneh Taghian, and Fatemeh Hajibabaie. (2022). Cross Brain-Gut Analysis Highlighted Hub Genes and LncRNAs Networks Differentially Modified During Leucine Consumption and Endurance Exercise in Mice with Depression Like Behaviors. *Molecular Neurobiology*, IF: 5.5.

[8] Navid Abedpoor, Iman Niktab, Mohammad-Hossein Beigi, Masoud Baghi, Fahimeh Arzande, Naeimeh Rezaei, Mohammad-Sajad Zare, Timothy L. Megraw,

Hoi-Ying Holman, Amirkianoosh Kiani, Farzad Seyed Forootan, Hossein Baharvand, Mohammad Hossein Nasr Esfahani, and Kamran Ghaedi (2022). Exercise facilitates the browning of fat tissue by up-regulating Irisin receptors. (Submitted).

[9] Maryam Haghparast Azad, Iman Niktab, Shaghayegh Dastjerdi, Navid Abedpoor, Golbarg Rahimi, Zahra Safaeinejad, Maryam Peymani, Farzad Seyed Forootan, Majid Asadi-Shekaari, Mohammad Hossein Nasr Esfahani and Kamran Ghaedi. (2022). The combination of endurance exercise and SGTC (Salvia–Ginseng–Trigonella–Cinnamon) ameliorate mitochondrial markers' overexpression with sufficient ATP production in the skeletal muscle of mice fed AGEs-rich high-fat diet. *Nutrition & Metabolism.*

[10] Golnaz Pakravan, Maryam Peymani, Navid Abedpoor, Zahra Safaeinejad, Mehrdad Yadegari, Maryam Derakhshan, Mohammad Hossein Nasr Esfahani, and Kamran Ghaedi. (2022). Antiapoptotic and anti-inflammatory effects of Pparγ agonist, pioglitazone, reversed Dox-induced cardiotoxicity through mediating of miR-130a downregulation in C57BL/6 mice. *Journal of Biochemical and Molecular Toxicology.*

[11] Fatemeh Azizian-Farsani, Navid Abedpoor, Maryam Derakhshan, Mohammad Hossein Nasr-Esfahani, Mohammad Hasan Sheikhha, and Kamran Ghaedi. (2022). Protective Effects of the Combination of the Herbal Compound Against Inflammation Related to Obesity and Colitis Induced by Diet in Mice. *Iranian Journal of Diabetes and Obesity.*

[12] Global multi-stakeholder endorsement of the MAFLD definition. (2022). *The Lancet Gastroenterology & Hepatology.* IF: 200.

[13] Fatemeh Hajibabaie, Faranak Aali, and Navid Abedpoor*. Pathomechanisms of non-coding RNAs and hub genes related to the oxidative stress in diabetic complications. (2022). F1000Research.

[14] Fatemeh Hajibabaie, Navid Abedpoor*, Kamran Safavi, and Taghian, F. (2022). Natural remedies medicine derived from flaxseed (secoisolariciresinol diglucoside, lignans, and α-linolenic acid) improve network targeting efficiency of diabetic heart conditions based on computational chemistry techniques and pharmacophore modeling. *Journal of food biochemistry.*

[15] Fatemeh Hajibabaie, Navid Abedpoor*, Farzaneh Taghian, and Kamran Safavi. (2022). A cocktail of polyherbal bioactive compounds and regular mobility training as senolytic approaches in age-dependent Alzheimer's: The in-silico analysis, lifestyle intervention in old age. *Journal of Molecular Neuroscience.*

Name: Navid Abedpoor
Affiliation: Department of Sports Physiology, Faculty of Sports Sciences, Isfahan (Khorasgan) Branch, Islamic Azad University, Isfahan, Iran.
Department of Physiology, Medicinal Plants Research Center, Isfahan (Khorasgan) Branch, Islamic Azad University, Isfahan, Iran.

Education: Sport sciences

Business Address: Isfahan (Khorasgan) Branch, Islamic Azad University, Isfahan, Iran.

Research and Professional Experience: Inflammation, Oxidative Stress, Non coding RNAs, Cancer, Lifestyles

Professional Appointments: Researcher

Honors:
1. **Won the Young Scientist** Award in "International Scientist Awards on Engineering, Science, and Medicine." 2020.
2. **Best Poster** in Royan International twin congress, Reproductive Biomedicine & Stem Cell. Protective approaches of Fraxinus excelsior compounds on the Implantation based infertility via bioinformatics and chemoinformatic analysis.

Publications from the Last 3 Years:

[1] Fatemeh Azizian-Farsani, Navid Abedpoor, Mohammad Hasan Sheikhha, Ali Osmay Gure, Mohammad Hossein Nasr Esfahani, Kamran Ghaedi. (2020). Receptor for advanced glycation end products acts as a fuel to colorectal cancer development. Frontiers in Oncology. IF: 6.5

[2] Fatemeh Azizian-Farsani, Marcin Osuchowski, Navid Abedpoor, Farzad Seyed Forootan, Maryam Derakhshan, Mohammad Hossein Nasr-Esfahani, Mohammad Hasan Sheikhha, Kamran Ghaedi. (2020). Anti-inflammatory and -apoptotic effects of an herbal extract on DSS-induced colitis in mice fed with high AGEs-fat diet. Scientific Reports. Nutrition & Metabolism. IF: 4.5

[3] Golbarg Rahimi; Salime Heydari; Bahare Rahimi; Navid Abedpoor; Iman Nicktab; Zahra Safaeinejad; Maryam Peymani; Farzad Seyed Forootan; Zahra Derakhshan; Mohammad Hossein Nasr Esfahani, Kamran Ghaedi. (2020). A combination of herbal compound (SPTC) along with exercise or metformin more efficiently alleviated diabetic complications through down-regulation of stress oxidative pathway upon activating Nrf2-Keap1 axis in AGEs rich diet-induced type 2 diabetic mice. Nutrition and metabolism.IF:4.5

[4] Fahimeh Akbarian, Mohsen Rahmani, Marziyeh Tavalaee, Navid Abedpoor, Mozhdeh Taki, Kamran Ghaedi, Mohammad Hossein Nasr-Esfahani. (2021). Effect of different high-fat and AGEs diets in obesity and diabetes-prone C57BL/6 mice on sperm function. International Journal of Fertility and Sterility. IF:2.7

[5] Navid Abedpoor, Farzaneh Taghian, Fatemeh Hajibabaie. (2022). Physical activity ameliorates the function of organs via adipose tissue in metabolic diseases. Acta histochemical. IF: 2.7

[6] Fatemeh Hajibabaie, Navid Abedpoor, Nazanin Asareh, Mohammad Amin Tabatabaiefar, Ali Zarrabi and Laleh Shariati. (2022). A cocktail of microRNAs as an advance diagnostic signature in stomach-colorectal cancers hallmarks incidence: a systematic review. Personal Medicine.

[7] Navid Abedpoor, Farzaneh Taghian, Fatemeh Hajibabaie. (2022). Cross Brain-Gut Analysis Highlighted Hub Genes and LncRNAs Networks Differentially Modified During Leucine Consumption and Endurance Exercise in Mice with Depression Like Behaviors. Molecular Neurobiology. IF: 5.5

[8] Navid Abedpoor, Iman Niktab, Mohammad-Hossein Beigi, Masoud Baghi, Fahimeh Arzande, Naeimeh Rezaei, Mohammad-Sajad Zare, Timothy L. Megraw, Hoi-Ying Holman, Amirkianoosh Kiani, Farzad Seyed Forootan, Hossein Baharvand, Mohammad Hossein Nasr Esfahani, Kamran Ghaedi (2022). Exercise facilitates the browning of fat tissue by up-regulating Irisin receptors. (Submitted).

[9] Maryam Haghparast Azad, Iman Niktab, Shaghayegh Dastjerdi, Navid Abedpoor, Golbarg Rahimi, Zahra Safaeinejad, Maryam Peymani, Farzad Seyed Forootan, Majid Asadi-Shekaari, Mohammad Hossein Nasr Esfahani & Kamran Ghaedi. The combination of endurance exercise and SGTC (Salvia–Ginseng–Trigonella–Cinnamon) ameliorate mitochondrial markers' overexpression with sufficient ATP production in the skeletal muscle of mice fed AGEs-rich high-fat diet. (2022). Nutrition & Metabolism.

[10] Golnaz Pakravan, Maryam Peymani, Navid Abedpoor, Zahra Safaeinejad, Mehrdad Yadegari, Maryam Derakhshan, Mohammad Hossein Nasr Esfahani, Kamran Ghaedi. Antiapoptotic and anti-inflammatory effects of Pparγ agonist, pioglitazone, reversed Dox-induced cardiotoxicity through mediating of miR-130a downregulation in C57BL/6 mice. (2022). Journal of Biochemical and Molecular Toxicology.

[11] Fatemeh Azizian-Farsani, Navid Abedpoor, Maryam Derakhshan, Mohammad Hossein Nasr-Esfahani, Mohammad Hasan Sheikhha, Kamran Ghaedi. Protective Effects of the Combination of the Herbal Compound Against Inflammation Related to Obesity and Colitis Induced by Diet in Mice. (2022). Iranian Journal of Diabetes and Obesity.

[12] Global multi-stakeholder endorsement of the MAFLD definition. (2022). The Lancet Gastroenterology & Hepatology. IF: 200

[13] Fatemeh Hajibabaie, Faranak Aali, Navid Abedpoor*. Pathomechanisms of non-coding RNAs and hub genes related to the oxidative stress in diabetic complications. (2022). F1000Research.

[14] Fatemeh Hajibabaie, Navid Abedpoor*, Kamran Safavi, F. Taghian. Natural remedies medicine derived from flaxseed (secoisolariciresinol diglucoside, lignans, and α-linolenic acid) improve network targeting efficiency of diabetic heart conditions based on computational chemistry techniques and pharmacophore modeling. (2022). Journal of food biochemistry.

[15] Fatemeh Hajibabaie, Navid Abedpoor*, Farzaneh Taghian, Kamran Safavi. A cocktail of polyherbal bioactive compounds and regular mobility training as senolytic approaches in age-dependent Alzheimer's: The in-silico analysis, lifestyle intervention in old age. (2022). Journal of Molecular Neuroscience.

Chapter 2

The Influence of Linolenic Acid on the Immune Response

Elina Kaviani[1]
Kamran Safavi[2]
Fatemeh Hajibabaie[3,4]
and Navid Abedpoor[4,*]

[1]Isfahan Endocrine and Metabolism Research Center,
Isfahan University of Medical Sciences, Isfahan, Iran
[2]Department of Plant Biotechnology, Medicinal Plants Research Centre,
Isfahan (Khorasgan) Branch, Islamic Azad University, Isfahan, Iran
[3]Department of Biology, Faculty of Basic Sciences, Shahrekord Branch,
Islamic Azad University, Shahrekord, Iran
[4]Department of Physiology, Medicinal Plants Research Center,
Isfahan (Khorasgan) Branch, Islamic Azad University, Isfahan, Iran

Abstract

Alpha-linolenic acid (ALA; 18:3(ω-3)) is an essential fatty acid whose dominant source is seed oils, including flax and chia. It can be partially converted into the other long-chain ω-3 PUFAs (including eicosapentaenoic acid (EPA; 20:5) and docosahexaenoic acid (DHA; 20:6) by multiple elongases, desaturases, and β-oxidases. However, EPA and DHA can typically be provided from fish oil. In recent decades, the impact of the omega-3 family (ALA, EPA, and DHA) on the immune system has received much attention due to its essential regulatory functions. They can incorporate into cell membranes and subsequently influence cell membrane properties, such as membrane fluidity or

---

[*] Corresponding Author's Email: Abedpoor.navid@yahoo.com, Abedpoor.navid@gmail.com.

In: Properties and Uses of Linolenic Acid
Editor: Calvin S. Willmon
ISBN: 979-8-89113-201-6
© 2023 Nova Science Publishers, Inc.

complex protein assembly in lipid rafts. Furthermore, long-chain ω-3 PUFAs molecules and their metabolites influence the effector and regulatory functions of innate and adaptive immune cells as signaling molecules. These regulations are usually the result of the secretion of cytokines and chemokines by cells, which attract immune cells from the circulation into inflammatory sites and, ultimately, determine activation or suppression of them. ALA and its derivatives suppress the production of pro-inflammatory cytokines (TNF-α, 37 IL-1β) and linoleic acid (ω-6)-derived eicosanoids (thromboxane B2 and prostaglandin E2). Also, many studies demonstrated that ALA has anti-inflammatory and antioxidant effects by increasing the expression level of nuclear factor erythroid 2–related factor 2 (Nrf2, a regulatory protein for cellular resistance to oxidants). Indeed, the anti-inflammatory and antioxidant properties of the omega-3 family can also prevent chronic-inflammatory and cancer. This chapter describes the effect of linolenic acid and its derivatives on the innate and adaptive immune response.

Keywords: Alpha-linolenic acid, immune system, cytokines and chemokines, innate and adaptive immune response

Introduction

The immune system is a complex network of cells, organs, tissue, proteins, and chemical mediators that protects the body from infections and other diseases (Innes and Calder 2018). Coordination and regulation of the different immune cells' activity are vital for generating an effective immune response (Figure 1). Innate and adaptive immunity are the two primary arms of the immune system. Macrophages, neutrophils, eosinophils, basophils, mast cells, natural killer cells, and dendritic cells are cells from the innate immune system and the first cellular line of defense against foreign antigens (Vancolen, Sébire, and Robaire 2023). They have an antigen-independent (non-specific) response immediately activated after encountering the antigen. Innate immunity cannot develop an immunologic memory system, hence being unable to recognize or defend against the same pathogen in the future (Hira 2022).

In contrast with the innate system, adaptive immunity relies on fewer types of cells with slower and more specific responses: B cells and T cells. Indeed, the hallmark of adaptive immunity is the memory system expansion against pathogens after a first confrontation (Chen and Colonna 2022). Cell-secreted mediators such as cytokines, chemokines, and other signaling molecules can regulate the cross-talk between innate and adaptive systems. For example, during the early response to bacterial infection, key inflammatory cytokines such as necrosis factor (TNF), interleukin 1 (IL-1), and interleukin 6 (IL-6) are secreted to initiate cell recruitment and local inflammation for clearance of the pathogen. They can also stimulate the growth and differentiation of lymphocytes through the mobilization and activation of antigen-presenting cells (APCs) (Sharma et al., 2022).

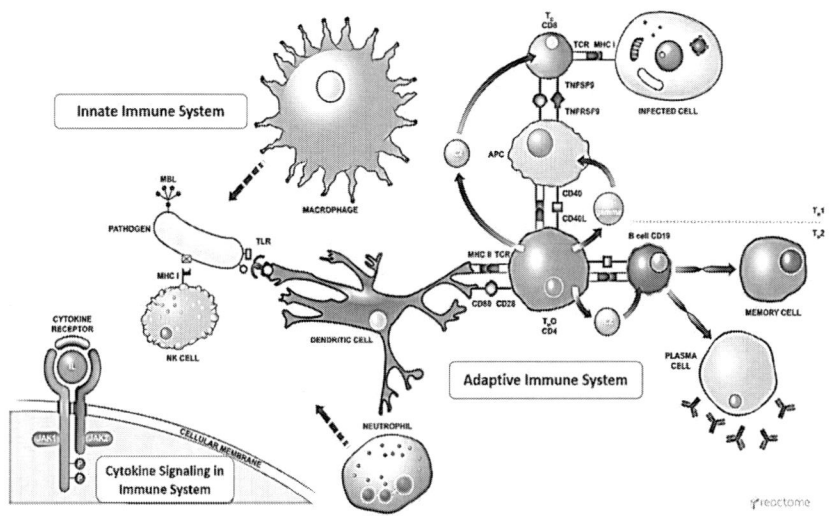

Figure 1. The innate and adaptive immune mechanisms collaborate to fight and eliminate pathogens.

Notably, ω-3 PUFAs, as a constitutive part of the cellular membrane, have immune-regulatory properties and thus exert major alterations on cell activation (Snodgrass and Brüne 2019). Their incorporation into the membrane phospholipids can affect the function of transmembrane proteins and membrane biophysical parameters such as increasing membrane fluidity, a higher degree of hydration, decreasing thickness, and increasing bending elasticity of lipid bilayers (Figure 2) (I. Levental and Lyman 2023). Indeed, after consumption, ω-3 PUFAs compete with the ω-6 family for incorporation into the phospholipid bilayer of cell membranes. Given the fact that arachidonic acid (AA; 20:4 (ω-6) is the most important member of the ω-6 family in the cell membrane, and following stimulation, it is released and converted to cellular mediators such as thromboxanes, prostaglandins, and leukotrienes (K.R. Levental et al., 2020). Omega-3 PUFAs can directly impact

AA metabolism by displacing it from membranes as well as competing for the recruitment of the enzymes that catalyze the biosynthesis of AA-derived mediators. Thus, the ω-3 PUFAs EPA and DHA can reduce AA-mediated inflammatory responses (Innes and Calder 2018). In Figure 3, the metabolic pathway for omega-3 and omega-6 can be observed (Figure 3).

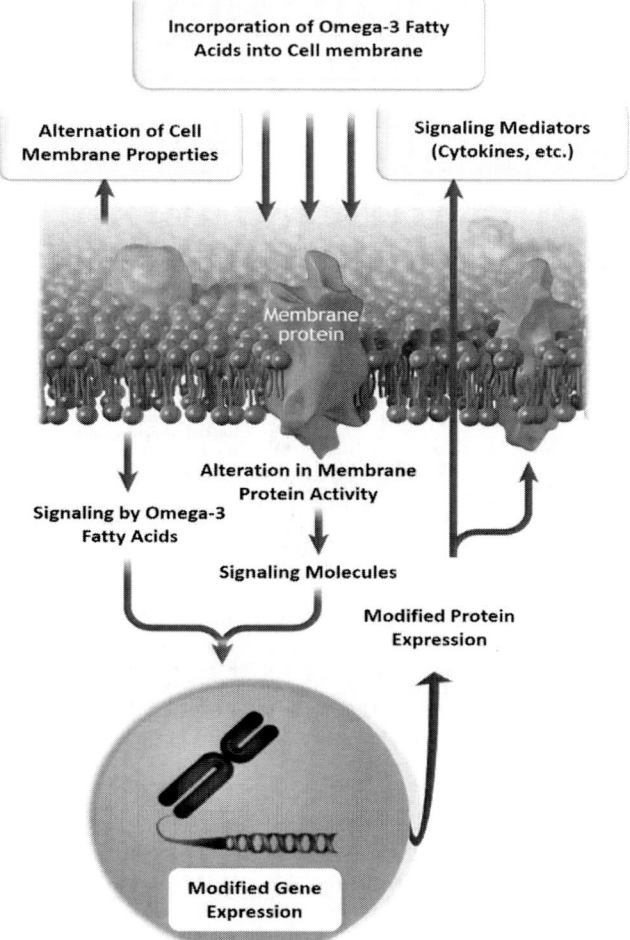

Figure 2. Incorporation of omega-3 fatty acids into the cell membrane. Omega-3 fatty acids can modulate membrane fluidity and membrane protein activity, and subsequently, alter protein and gene expression.

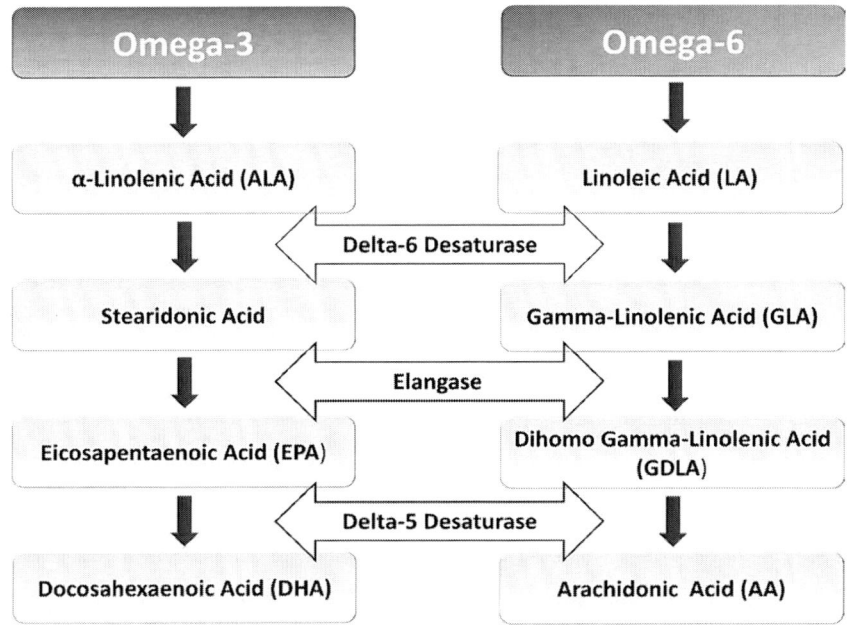

Figure 3. Metabolic pathway of omega-3 and omega-6 fatty acids.

Additionally, EPA and DHA active peroxisome proliferator-activated receptor (PPAR)-γ and G-protein coupled receptor (GPR) 120, and subsequently inhibit the activation of pro-inflammatory transcription factor NF-κB (Calder 2019). Accordingly, ω-3 PUFAs directly affect intracellular signaling processes, gene expression, and the production of mediators (Jarc and Petan 2020).

Omega-3 PUFAs and Innate Immune Responses

Epithelium

The epithelium is a biochemical and physical mucosal barrier between the environment and the internal milieu of the body (Ornelas et al., 2022). This barrier comprises a continuous layer of epithelial cells that is sealed between them by tight junctions. Indeed, tight junctions contain a variety of transmembrane proteins (occludin and claudin), junctional adhesion molecules (JAMs), intracellular scaffold proteins (zonula occludens (ZO)-1) that play a critical role in regulating and selecting solutes moving across the epithelium (Figure 4). As the first line of protection, epithelial cells are present in the gut, lungs, and skin. They express pattern recognition receptors, such as Toll-like receptors (TLRs), to recognize pathogen-associated molecular patterns (PAMPs) and danger-associated molecular patterns (DAMPs), which results in the secretion of pro and anti-inflammatory cytokines, chemokines, and mediators (Naylor et al., 2019; Caballero-Herrero et al., 2023). In addition, these interactions can directly or indirectly affect the response of other innate and adaptive immune cells (Ornelas et al., 2022).

One of the vital epithelium functions is the regulation of permeability, which can be disturbed by factors such as the response to the increased presence of inflammatory cytokines. Indeed, these cytokines induce the degradation or endocytosis of tight junction protein (Cunningham and Turner 2012). Many in vitro studies have shown that EPA and DHA have the potential to improve intestinal barrier integrity and even restore membrane permeability after inflammatory cytokines-mediated disruption (Q. Li et al., 2008; Willemsen et al., 2008; Beguin et al., 2013). As mentioned earlier, ω-3 PUFAs exert anti-inflammatory effects and decrease the secretion of inflammatory cytokines. EPA and DHA stimulate GPR120, which subsequently, cytosolic calcium is accumulated. Then, the ERK cascade is induced, and NF-κB is inhibited (Hirasawa et al., 2005; Mobraten et al., 2013). In vivo studies have also confirmed the protective effect of ω-3 PUFAs in the gut. For example, the DHA-treated IL-10⁻ / ⁻ mice (model of chronic colitis) improve the intestinal epithelial barrier function by reducing permeability, increasing transepithelial electrical resistance (TEER), and up-regulation of occludin and ZO-1 expression. Additionally, DHA reduced pro-inflammatory cytokines such as IL-17, TNF-α, and IFN-γ (Zhao et al., 2015). This founding was consistent with the results obtained from the 2005 study on the ω-3 PUFA-treated SCID mice model of colitis (Whiting, Bland, and Tarlton 2005). Another mechanism of the ω-3 PUFA's protective impacts in the intestine is the reduction of oxidative stress (Hassan et al., 2010).

Anti-inflammatory effects of DHA have been proven by reducing IL-6 and CXCL10 secretion from the airway epithelium cells after rhinovirus infection (Saedisomeolia et al., 2008). Furthermore, according to the literature, DHA can enhance the healing capacity by up-regulation of Amphiregulin (AREG), an epidermal growth factor family member, after inflammatory dust-induced lung injury (Nordgren et al., 2018).

Briefly, ω-3 PUFAs exert their effects on epithelial cells by activating transmembrane and nuclear receptors, as well as by modifying the ω-6/ω-3

PUFA content of cell membranes. In the presence of underlying inflammation, they have the potential to improve disturbed barrier function and reduce the production and secretion of pro-inflammatory mediators (D'Angelo, Motti, and Meccariello 2020).

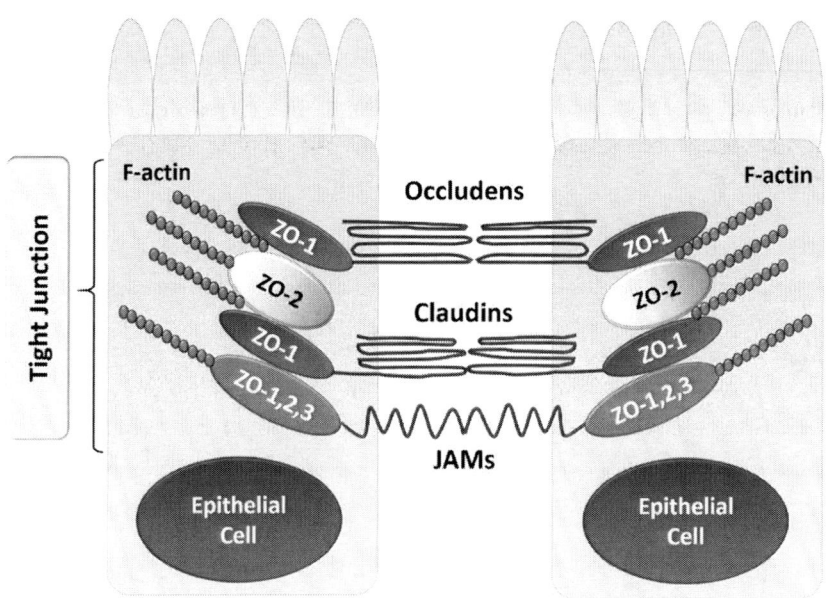

Figure 4. Schematic representation of epithelial tight junction proteins and their location.

Macrophages

Macrophages are an important population of tissue-resident effector cells of innate immunity. They phagocyte and process the foreign antigens, diseased and damaged cells, and regulatory and recruit other macrophages during an inflammatory response. Indeed, they recognize the PAMPs with their TLRs. During the elimination process, they ingest the invasive pathogens and produce the reactive oxygen species (Yan et al.). Their population is highly heterogeneous, which has the potential to change rapidly in response to various microenvironmental signals. The basis on of their function and activation are classified into three groups: classically activated macrophages (M1 macrophages), alternatively activated macrophages (M2 macrophages), and regulatory macrophages (Mregs). M1 macrophages produce proinflammatory cytokines and mediate resistance to various bacteria, protozoa, and viruses as well as, anti-tumor responses. M2 macrophages are involved in the anti-inflammatory cytokine secretion and the wound-healing process. Mregs produce and secrete a high level of anti-inflammatory cytokine IL-10 during the response to Fc receptor-γ ligation.

Omega-3 PUFAs can modulate three main properties of macrophage functions, including the production and secretion of cytokines and chemokines, phagocytic activity, and macrophage polarization. The ω-3 PUFAs alter the gene expression profile in macrophages, although the effect of each member is not identical. For example, most EPA and DHA affect expression patterns of genes related to cell cycle regulation and immune response, respectively. According to previous studies, the anti-inflammatory potency of DHA is higher than EPA, and intriguingly, the combination of DHA+EPA can enhance immunomodulatory response as compared to each of them alone in macrophages (Weldon et al., 2007; Allam-Ndoul et al., 2017).

Following an LPS infection and subsequent LPS/TLR4 interaction, an NF-κB-dependent inflammatory cascade is induced, producing proinflammatory cytokines. Interestingly, ω-3 PUFAs suppress the signaling pathway of TLR4, which subsequently inhibits LPS-induced NFκB activation and cyclooxygenase (COX-2) expression and as well as prostaglandin release in the macrophage cells (Honda et al., 2015; J.Y. Lee et al., 2003; J.Y. Lee et al., 2001). Additionally, omega-3 fatty acids can reduce ROS and NO production (Kumar et al., 2016; Sung et al., 2017), associated with decreased activation of inflammatory transcription factors. On the other hand, ω-3 PUFAs have the potential to inhibit the NLRP3 inflammasome (a multiprotein complex with a central role in causing inflammation) mediated by PPARγ and GPR120/GRP40 signaling (Yan et al., 2013; K.R. Lee, Midgette, and Shah 2019).

The anti-inflammatory properties of ω-3 PUFAs on macrophages decrease the secretion of IL-1β, TNF-α, IL-6 and subsequent M1 macrophage polarization. Indeed, ω-3 PUFAs stimulate the expression of M2 markers such as IL-10, YM-1, arginase, Clec7a and MMR, indicating an increased M2 macrophage polarization (Talukdar et al., 2010). The treatment with ω-3 fatty acids enhances the phagocytic capacity and microbicidal activity of macrophages as well as decreases apoptosis (Saini, Harjai, and Chhibber 2013; Adolph, Fuhrmann, and Schumann 2012). Increasing the phagocytic capacity may be related to changing the cellular membrane structure through omega-3 incorporation (Hellwing et al., 2018).

Neutrophil

Neutrophils make up the most abundant population of white blood cells in human, which are physiologically distributed in two pools: a circulating neutrophil pool (CNP) and a marginated neutrophil pool (MNP). When the infection enters the body, they migrate to the infection site and phagocyte them by a phagocytic vacuole. This vacuole is destroyed by an acidic pH and degradation enzymes. Neutrophils also maintain host integrity by mobilizing their granules, generating neutrophil extracellular traps, and cytokine production. Additionally, neutrophils play an important role in the activation of the adaptive immune system (Omman and Kini 2019).

Omega-3 fatty acids, and their derivatives, affect migration, phagocytic capacity, and ROS and cytokine production of neutrophils (Gutiérrez, Svahn, and Johansson 2019; Radzikowska et al., 2019). Endothelial cells provide neutrophil migration conditions through prostaglandin D2 (PGD2) production, as a metabolite derived of AA. Indeed, produced PGD2 binds to DP1 receptor on neutrophils and subsequent it contributes to neutrophil adhesion and migration. Interestingly, EPA treated-endothelial cells produce an alternative eicosanoid, prostaglandin D3, which inhibits migration of neutrophils (Tull et al., 2009). According to previous studies, Resolvin D1 (RvD1), a metabolite of ω-3 PUFAs derived from DHA, decrease neutrophil migration by reducing in actin polymerization (Krishnamoorthy et al., 2010). However, an in vivo study demonstrated that the ω-3 PUFAs effect on neutrophil migration is a time dependent factor (Arnardottir, Freysdottir, and Hardardottir 2013).

DHA can significantly enhance the phagocytic capacity and antifungal properties of the neutrophils. As mentioned earlier, the generation and release of ROS is an important defense mechanism for neutrophils, on which omega-3 fatty acids can have different effects, depending on the animal model used and age (in humans) (Gutiérrez, Svahn, and Johansson 2019). For example, DHA and especially EPA increased ROS production in rat neutrophils, whereas DHA decreased its amount and EPA did not affect it in the goat. In a human study, ROS production has increased in treated volunteers with the combination of EPA (26%) + DHA (54%) for two months. However, in another experiment on humans, EPA supplementation has decreased ROS production in older men but it had no effect on younger men (Gutiérrez, Svahn, and Johansson 2019).

Dendritic Cells

Dendritic cells (DCs) are considered as one of the most important antigen-presenting cells that play a critical role in the initiation and regulation of adaptive immune responses. In addition, they establish a strong link between innate and adaptive immunity. They can trap the antigen, then process and provide it on the cell surface to the T cells (He et al., 2019). Also, they regulate immune response by the production and secretion of IL-6, IL-12, TNF-α, IFN-α, and chemokines.

Omega-3 fatty acids can influence the DC functions and subsequently T cell response. Treatment of DC with DHA lead to reduce their maturation, co-stimulatory molecules expression, and the secretion of pro-inflammatory cytokines IL-12p70, IL-23, and IL-6. Furthermore, DHA-treated DCs have been reported to be a weak stimulator of antigen-specific T cell proliferation as well as Th1/Th17 differentiation. (Kong, Yen, and Ganea 2011). According to evidences found, DHA and EPA have the potential to reduce the MHC II and co-stimulatory molecules in the human and mouse DC surface (H. Wang et al., 2007; Kong et al., 2010; Zapata-Gonzalez et al., 2008; Zeyda et al., 2005).

Omega-3 PUFAs and Adaptive Immune Responses

T Cells

T cells have a prominent role in the immune system. T progenitor cells migrate from bone marrow to the thymus, where they mature. Mature T cells have a unique TCR receptor that that recognizes the antigens presented. While in the thymus, they express CD4 and CD8 receptors and are differentiated into two main subsets with different immune properties: $CD4^+$ and $CD8^+$ T cells. $CD4^+$ T cells (or helper T cells) help fight against bacterial infections, and $CD8^+$ T (or cytotoxic T lymphocytes) cells play a major role in the immune defense against viral infections. Helper T cells (Th) are classified into several subsets such as Th1, Th2, Th17, and Th22. The cytokine signature: for Th1 is IFN-γ, for Th2 is IL-4, for Th17 is IL-17A, IL17-F, IL-21, and IL-22 and for Th22 is IL-22. Another subset of $CD4^+$ T cells is regulatory T cells (Tregs) which help immune homeostasis by suppressing immune cells' response. For the regulation of immune cells, Tregs produce and secret the immunomodulatory cytokines (IL-10 and TGF-β). Pro-inflammatory T cells include the CD8⁺ T cells and the CD4⁺ Th1 and Th17 cells (Raphael et al., 2015).

T cell activations require the engagement of TCR by APCs, such as DCs or macrophages. Thereby, the effect of ω-3 fatty acids on APCs has an undeniable role in modulating T cell activation indirectly. Additionally, ω-3 fatty acids can directly influence T cell activation. Evidence from several studies has demonstrated that ω-3 fatty acids exert immunosuppressive effects on T cell functions. Indeed, ω-3 fatty acids supplementation has exerted promising effects in T cell-mediated autoimmune diseases such as asthma and hepatitis (Y. Li et al., 2016; Farjadian et al., 2016).

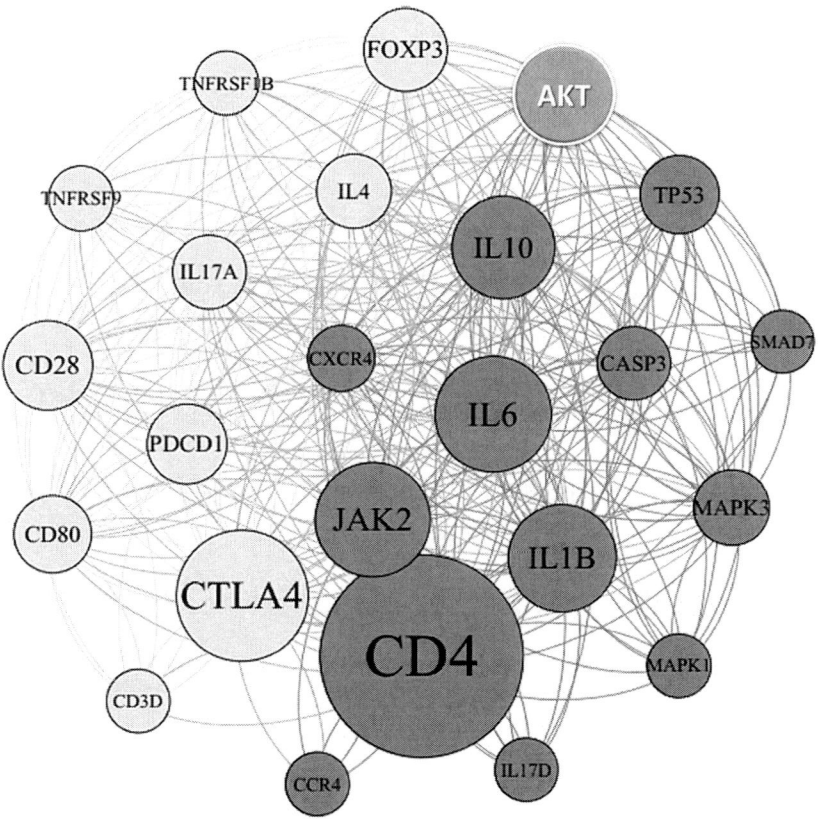

Figure 6. Protein-protein interaction analysis of PKB/Akt network. A slight alter in the PKB/Akt pathway can affect many signaling pathways of immune cells.

According to reports and evidences, EPA and DHA can reduce T cell proliferation by disrupting the autocrine IL-2 pathway (Jolly, McMurray, and Chapkin 1998; McMurray, Jolly, and Chapkin 2000). Moreover, omega-3 fatty acids could increase the FoxP3 expression and differentiation of Tregs (through the expansion of the M2 macrophage population as a direct inducer for Treg differentiation) (Kim et al., 2018; Han et al., 2015). This enriched Treg population leads to increase expression of immunosuppressive cytokines (TGF-β and IL-10) and ultimately reduces the secretion of pro-inflammatory cytokines (IFN-γ, IL-6, IL-23, and IL-17) from $CD4^+$ T cells and macrophages (Han et al., 2015; Yessoufou et al., 2009). Additionally, multiple studies have shown that mega-3 fatty acids reduce the $CD4^+$ T cells' differentiation into Th17 and subsequently decrease the IL-17 secretion (Jeffery et al., 2017; Chiurchiù et al., 2016; Farjadian et al., 2016; Allen et al., 2014; Shoda et al., 2015).

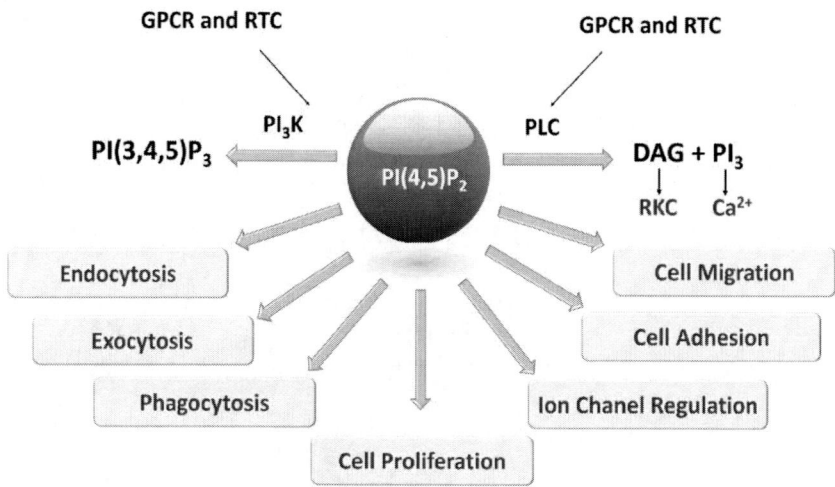

Figure 7. Multiple functions of $PI4,5P_2$ in the cytosol and plasma membrane. It is a key second messenger and a substrate for important signaling proteins including phosphoinositide-specific (PLC) and phosphoinositide 3-kinase (PI3K). This molecule regulates a variety of cellular activities.

Omega-3 fatty acids decrease the phosphorylation of Janus kinase 1 (JAK1), JAK3, STAT5, ERK1/2, and Protein Kinase B (Hassan et al., Gorjao et al., 2007), which subsequently alter many signal transductions pathways (Figure 6). Following these alternations, the cytotoxic T-lymphocyte-associated protein 4 (CTLA-4) expression increases in T cells (Ly et al., 2006;

Yessoufou et al., 2009). CTLA-4 is an inhibitory receptor that transduces the immunoinhibitory signals in T cells. Indeed, it competes with C28 (as a costimulatory receptor) for binding to CD80 or CD86 and ultimately preventing T cell stimulation and promoting T cell exhaustion. The membrane incorporation of omega-3 PUFA disrupts the spatial organization of phosphatidylinositol 4,5-bisphosphate (PI(4,5)P$_2$) and thus prevents the essential signals for CD4$^+$ T cells proliferation (Hou et al., 2016) (Figure 7). According to evidences, omega-3 PUFA can induce the apoptosis by enhancing the ROS production in a dose-dependent manner (Terada et al., 2001; Cury-Boaventura et al., 2006; Zurier et al., 1999).

B Cells

Besides T cells, B cells are a subtype of lymphocytes that play a crucial role in adaptive immunity. They mediate the production and secretion the antigen-specific immunoglobulin (Ig) and function at the center of the humoral immune system. Additionally, they regulate the immune response through antigen presentation and the production of cytokines, chemokines, and other mediators (Chistiakov, Orekhov, and Bobryshev 2016).

B cells are derived from hematopoietic stem cells (HSCs) in the bone marrow. During B cell development from HSCs, Ig heavy chain rearrangement first occurs in the pro-B cells and then Pro-B cells become pre-B cells by the initiation of Ig light chain immunoglobulin rearrangement (Y. Wang et al., 2020). After a complete immunoglobulin rearrangement and generation of functional BCR, as well as testing for autoreactivity, they migrate from bone marrow to the spleen. Immature cells are sensitive to out-reactivity. In the spleen, they are further differentiated into follicular B cells or marginal zone B cells through transitional B cells (transitional type 1 and type 2) stages. There are two major subsets of mature B cells: B1, and B2 cells. Conventional B2 cells communicating with T cells can provide high-affinity antibody responses and generate immunological memory (De Silva and Klein 2015). B1 cells produce natural antibodies and defense against non-protein antigens and microbial pathogens (Rothstein et al., 2013). According to researchers, in mice, omega-3 fatty acids altered the percentage of these subpopulations in different tissues. For example, a diet rich in omega-3 fatty acids (fish oil) decreases percentages of naïve B cells and mature B cells in the bone marrow but no changes in the amount of B1 or B2 cells in the peritoneal cavity, additionally, increases the numbers of transitional type 1 B cells in the spleen (Teague et al., 2013; Teague et al., 2016; Tomasdottir et al., 2014). Following B cell stimulation through interaction with immune cells or direct binding to PAMPs, activated B cells up-regulate the expression of several molecules, such as CD40, MCH II, CD86, and CD80 (Harwood and Batista 2009; Tarlinton 2019). DHA can reduce anti-CD-40-induced p50 nuclear translocation, STAT6 phosphorylation and IL-6 production in human B cells. Interestingly, they decrease the IgE production of human B cells, and subsequent allergic reactions by disrupting the CD40 and the IL-4 signaling pathway (Weise et al., 2011). In vitro, DHA and remarkably EPA have down-regulated the secretion of the key immunoregulatory cytokines IL-10, TNF-α, and INF-γ from the B lymphocyte cell line (Raji) (Verlengia et al., 2004). However, most evidence in the literature supported the negative role of omega-3 fatty acids in the B cell activation, there is still controversy in this field and a definite conclusion cannot be reached due to the limitations of the studies. For example, in vivo and ex vivo analysis has demonstrated that DHA can enhance B cell function by increasing the secretion of Th2-biasing cytokines (IL-5, IL-13, and IL-9) and IL-6 and TNF-α, CD40 expression, and cecal IgA (Gurzell et al., 2013). Another in vivo study has shown that an omega-3 fatty acids diet increases the CD69 expression and the production of IL-6 and IFNγ in B cells (Rockett et al., 2010).

Conclusion

Long-chain ω-3 PUFAs molecules and their metabolites influence the effector and regulatory functions of innate and adaptive immune cells as signaling molecules. These regulations are usually the result of the secretion of cytokines and chemokines by cells, which attract immune cells from the circulation into inflammatory sites and, ultimately, determine activation or suppression of them. ALA and its derivatives suppress the production of pro-inflammatory cytokines (TNF-α, 37 IL-1β) and linoleic acid (ω-6)-derived eicosanoids (thromboxane B2 and prostaglandin E2). Also, ALA has anti-inflammatory and antioxidant effects by increasing the expression level of nuclear factor erythroid 2–related factor 2 (Nrf2, a regulatory protein for cellular resistance to oxidants). Indeed, the anti-inflammatory and antioxidant properties of the omega-3 family can also prevent chronic-inflammatory and cancer.

References

Adolph, Stephanie, Herbert Fuhrmann, and Julia Schumann. 2012. "Unsaturated fatty acids promote the phagocytosis of P. aeruginosa and R. equi by RAW264. 7 macrophages." *Current microbiology* 65: 649-655.

Allam-Ndoul, B, F Guénard, O Barbier, and M C Vohl. 2017. "Effect of different concentrations of omega-3 fatty acids on stimulated THP-1 macrophages." *Genes & nutrition* 12: 1-8.

Allen, M Jeannie, Yang-Yi Fan, Jennifer M Monk, Tim Y Hou, Rola Barhoumi, David N McMurray, and Robert S Chapkin. 2014. "n–3 PUFAs reduce T-helper 17 cell differentiation by decreasing responsiveness to interleukin-6 in isolated mouse splenic CD4+ T cells." *The Journal of nutrition* 144 (8): 1306-1313.

Arnardottir, Hildur H, Jona Freysdottir, and Ingibjorg Hardardottir. 2013. "Dietary fish oil increases the proportion of a specific neutrophil subpopulation in blood and total neutrophils in peritoneum of mice following endotoxin-induced inflammation." *The Journal of Nutritional Biochemistry* 24 (1): 248-255.

Beguin, Pauline, Abdelmounaim Errachid, Yvan Larondelle, and Yves-Jacques Schneider. 2013. "Effect of polyunsaturated fatty acids on tight junctions in a model of the human intestinal epithelium under normal and inflammatory conditions." *Food & function* 4 (6): 923-931.

Caballero-Herrero, María José, Esther Jumilla, Manuel Buitrago-Ruiz, Graciela Valero-Navarro, and Santiago Cuevas. 2023. "Role of Damage-Associated Molecular Patterns (DAMPS) in the Postoperative Period after Colorectal Surgery." *International Journal of Molecular Sciences* 24 (4): 3862.

Calder, Philip C. 2019. "Is increasing microbiota diversity a novel anti-inflammatory action of marine n–3 fatty acids?" *The Journal of Nutrition* 149 (7): 1102-1104.

Chen, Yun, and Marco Colonna. 2022. "Spontaneous and induced adaptive immune responses in Alzheimer's disease: new insights into old observations." *Current Opinion in Immunology* 77: 102233.

Chistiakov, Dimitry A, Alexander N Orekhov, and Yuri V Bobryshev. 2016. "Immune-inflammatory responses in atherosclerosis: role of an adaptive immunity mainly driven by T and B cells." *Immunobiology* 221 (9): 1014-1033.

Chiurchiù, Valerio, Alessandro Leuti, Jesmond Dalli, Anders Jacobsson, Luca Battistini, Mauro Maccarrone, and Charles N Serhan. 2016. "Proresolving lipid mediators resolvin D1, resolvin D2, and maresin 1 are critical in modulating T cell responses." *Science translational medicine* 8 (353): 353ra111-353ra111.

Cunningham, Kevin E, and Jerrold R Turner. 2012. "Myosin light chain kinase: pulling the strings of epithelial tight junction function." *Annals of the New York Academy of Sciences* 1258 (1): 34-42.

Cury-Boaventura, Maria F, Renata Gorjão, Thais Martins De Lima, Philip Newsholme, and Rui Curi. 2006. "Comparative toxicity of oleic and linoleic acid on human lymphocytes." *Life sciences* 78 (13): 1448-1456.

D'Angelo, Stefania, Maria Letizia Motti, and Rosaria Meccariello. 2020. "ω-3 and ω-6 polyunsaturated fatty acids, obesity and cancer." *Nutrients* 12 (9): 2751.

De Silva, Nilushi S, and Ulf Klein. 2015. "Dynamics of B cells in germinal centres." *Nature reviews immunology* 15 (3): 137-148.

Farjadian, Shirin, Mozhgan Moghtaderi, Mehdi Kalani, Tahereh Gholami, and Saeed Hosseini Teshnizi. 2016. "Effects of omega-3 fatty acids on serum levels of T-helper cytokines in children with asthma." *Cytokine* 85: 61-66.

Gorjao, Renata, Sandro Massao Hirabara, Thaís Martins de Lima, Maria Fernanda Cury-Boaventura, and Rui Curi. 2007. "Regulation of interleukin-2 signaling by fatty acids in human lymphocytes." *Journal of lipid research* 48 (9): 2009-2019.

Gurzell, Eric A, Heather Teague, Mitchel Harris, Jonathan Clinthorne, Saame Raza Shaikh, and Jenifer I Fenton. 2013. "DHA-enriched fish oil targets B cell lipid microdomains and enhances ex vivo and in vivo B cell function." *Journal of leukocyte biology* 93 (4): 463-470.

Gutiérrez, Saray, Sara L Svahn, and Maria E Johansson. 2019. "Effects of omega-3 fatty acids on immune cells." *International journal of molecular sciences* 20 (20): 5028.

Han, Sang-Chul, Dong-Hwan Koo, Na-Jin Kang, Weon-Jong Yoon, Gyeoung-Jin Kang, Hee-Kyoung Kang, and Eun-Sook Yoo. 2015. "Docosahexaenoic acid alleviates atopic dermatitis by generating Tregs and IL-10/TGF-β-modified macrophages via a TGF-β-dependent mechanism." *Journal of Investigative Dermatology* 135 (6): 1556-1564.

Harwood, Naomi E, and Facundo D Batista. 2009. "Early events in B cell activation." *Annual review of immunology* 28: 185-210.

Hassan, Aktham, Ayman Ibrahim, Khaly Mbodji, Moïse Coëffier, Frédéric Ziegler, Frédéric Bounoure, Jean-Michel Chardigny, Mohamed Skiba, Guillaume Savoye, and Pierre Déchelotte. 2010. "An α-linolenic acid-rich formula reduces oxidative stress and inflammation by regulating NF-κB in rats with TNBS-induced colitis." *The Journal of nutrition* 140 (10): 1714-1721.

He, Zhimin, Xinyi Zhu, Zhen Shi, Tao Wu, and Li Wu. 2019. "Metabolic regulation of dendritic cell differentiation." *Frontiers in Immunology* 10: 410.

Hellwing, Christine, Feven Tigistu-Sahle, Herbert Fuhrmann, Reijo Käkelä, and Julia Schumann. 2018. "Lipid composition of membrane microdomains isolated detergent-free from PUFA supplemented RAW264. 7 macrophages." *Journal of cellular physiology* 233 (3): 2602-2612.

Hira, Princy. 2022. "Overview of Immune System." In *An Interplay of Cellular and Molecular Components of Immunology*, 1-26. CRC Press.

Hirasawa, Akira, Keiko Tsumaya, Takeo Awaji, Susumu Katsuma, Tetsuya Adachi, Masateru Yamada, Yukihiko Sugimoto, Shunichi Miyazaki, and Gozoh Tsujimoto. 2005. "Free fatty acids regulate gut incretin glucagon-like peptide-1 secretion through GPR120." *Nature medicine* 11 (1): 90-94.

Honda, Kaori L, Stefania Lamon-Fava, Nirupa R Matthan, Dayong Wu, and Alice H Lichtenstein. 2015. "EPA and DHA exposure alters the inflammatory response but not the surface expression of Toll-like receptor 4 in macrophages." *Lipids* 50: 121-129.

Hou, Tim Y, Rola Barhoumi, Yang-Yi Fan, Gonzalo M Rivera, Rami N Hannoush, David N McMurray, and Robert S Chapkin. 2016. "n-3 polyunsaturated fatty acids suppress CD4+ T cell proliferation by altering phosphatidylinositol-(4, 5)-bisphosphate [PI (4,

5) P2] organization." *Biochimica et Biophysica Acta (BBA)-Biomembranes* 1858 (1): 85-96.
Innes, Jacqueline K, and Philip C Calder. 2018. "Omega-6 fatty acids and inflammation." *Prostaglandins, Leukotrienes and Essential Fatty Acids* 132: 41-48.
Jarc, Eva, and Toni Petan. 2020. "A twist of FATe: Lipid droplets and inflammatory lipid mediators." *Biochimie* 169: 69-87.
Jeffery, Louisa, Helena L Fisk, Philip C Calder, Andrew Filer, Karim Raza, Christopher D Buckley, Iain McInnes, Peter C Taylor, and Benjamin A Fisher. 2017. "Plasma levels of eicosapentaenoic acid are associated with anti-TNF responsiveness in rheumatoid arthritis and inhibit the etanercept-driven rise in Th17 cell differentiation in vitro." *The Journal of rheumatology* 44 (6): 748-756.
Jolly, C A, DN McMurray, and RS Chapkin. 1998. "Effect of dietary n-3 fatty acids on interleukin-2 and interleukin-2 receptor α expression in activated murine lymphocytes." *Prostaglandins, leukotrienes and essential fatty acids* 58 (4): 289-293.
Kim, Ji Young, Kyu Lim, Kyung Hee Kim, Jin Hyun Kim, Jin Sun Choi, and Seung-Cheol Shim. 2018. "N-3 polyunsaturated fatty acids restore Th17 and Treg balance in collagen antibody-induced arthritis." *PLoS One* 13 (3): e0194331.
Kong, Weimin, Jui-Hung Yen, and Doina Ganea. 2011. "Docosahexaenoic acid prevents dendritic cell maturation, inhibits antigen-specific Th1/Th17 differentiation and suppresses experimental autoimmune encephalomyelitis." *Brain, behavior, and immunity* 25 (5): 872-882.
Kong, Weimin, Jui-Hung Yen, Evros Vassiliou, Sabina Adhikary, Miguel G Toscano, and Doina Ganea. 2010. "Docosahexaenoic acid prevents dendritic cell maturation and in vitro and in vivo expression of the IL-12 cytokine family." *Lipids in health and disease* 9 (1): 1-10.
Krishnamoorthy, Sriram, Antonio Recchiuti, Nan Chiang, Stephanie Yacoubian, Chih-Hao Lee, Rong Yang, Nicos A Petasis, and Charles N Serhan. 2010. "Resolvin D1 binds human phagocytes with evidence for proresolving receptors." *Proceedings of the National Academy of Sciences* 107 (4): 1660-1665.
Kumar, Naresh, Geetika Gupta, Kotha Anilkumar, Naireen Fatima, Roy Karnati, Gorla Venkateswara Reddy, Priyanka Voori Giri, and Pallu Reddanna. 2016. "15-Lipoxygenase metabolites of α-linolenic acid,[13-(S)-HPOTrE and 13-(S)-HOTrE], mediate anti-inflammatory effects by inactivating NLRP3 inflammasome." *Scientific reports* 6 (1): 31649.
Lee, Joo Y, Anthony Plakidas, Won H Lee, Anne Heikkinen, Prithiva Chanmugam, George Bray, and Daniel H Hwang. 2003. "Differential modulation of Toll-like receptors by fatty acids: preferential inhibition by n-3 polyunsaturated fatty acids." *Journal of lipid research* 44 (3): 479-486.
Lee, Joo Y, Kyung H Sohn, Sang H Rhee, and Daniel Hwang. 2001. "Saturated fatty acids, but not unsaturated fatty acids, induce the expression of cyclooxygenase-2 mediated through Toll-like receptor 4." *Journal of Biological Chemistry* 276 (20): 16683-16689.
Lee, Kailey Roberts, Yasmeen Midgette, and Rachana Shah. 2019. "Fish oil derived omega 3 fatty acids suppress adipose NLRP3 inflammasome signaling in human obesity." *Journal of the Endocrine Society* 3 (3): 504-515.

Levental, Ilya, and Ed Lyman. 2023. "Regulation of membrane protein structure and function by their lipid nano-environment." *Nature Reviews Molecular Cell Biology* 24 (2): 107-122.

Levental, Kandice R, Eric Malmberg, Jessica L Symons, Yang-Yi Fan, Robert S Chapkin, Robert Ernst, and Ilya Levental. 2020. "Lipidomic and biophysical homeostasis of mammalian membranes counteracts dietary lipid perturbations to maintain cellular fitness." *Nature communications* 11 (1): 1339.

Li, Qiurong, Qiang Zhang, Meng Wang, Sumin Zhao, Guowang Xu, and Jieshou Li. 2008. "n-3 polyunsaturated fatty acids prevent disruption of epithelial barrier function induced by proinflammatory cytokines." *Molecular immunology* 45 (5): 1356-1365.

Li, Yanli, Yuan Tang, Shoujie Wang, Jing Zhou, Jia Zhou, Xiao Lu, Xiaochun Bai, Xiang-Yang Wang, Zhengliang Chen, and Daming Zuo. 2016. "Endogenous n-3 polyunsaturated fatty acids attenuate T cell-mediated hepatitis via autophagy activation." *Frontiers in immunology* 7: 350.

Ly, Lan H, Roger Smith, Kirsten C Switzer, Robert S Chapkin, and David N McMurray. 2006. "Dietary eicosapentaenoic acid modulates CTLA-4 expression in murine CD4+ T-cells." *Prostaglandins, leukotrienes and essential fatty acids* 74 (1): 29-37.

McMurray, David N, Christopher A Jolly, and Robert S Chapkin. 2000. "Effects of dietary n-3 fatty acids on T cell activation and T cell receptor-mediated signaling in a murine model." *The Journal of infectious diseases* 182 (Supplement_1): S103-S107.

Mobraten, Kaia, Trude M Haug, Charlotte R Kleiveland, and Tor Lea. 2013. "Omega-3 and omega-6 PUFAs induce the same GPR120-mediated signalling events, but with different kinetics and intensity in Caco-2 cells." *Lipids in health and disease* 12 (1): 1-7.

Naylor, Aisling, Alan Hopkins, Natalie Hudson, and Matthew Campbell. 2019. "Tight junctions of the outer blood retina barrier." *International journal of molecular sciences* 21 (1): 211.

Nordgren, Tara M, Art J Heires, Kristina L Bailey, Dawn M Katafiasz, Myron L Toews, Christopher S Wichman, and Debra J Romberger. 2018. "Docosahexaenoic acid enhances amphiregulin-mediated bronchial epithelial cell repair processes following organic dust exposure." *American Journal of Physiology-Lung Cellular and Molecular Physiology* 314 (3): L421-L431.

Omman, Reeba A, and Ameet R Kini. 2019. *"Leukocyte development, kinetics, and functions."* Keohane EM, Otto CM, Walenga JM, Rodak's Hematology: Clinical Principles and Applications. 6th edn. Amsterdam, Netherlands: Elsevier: 117-135.

Ornelas, Alfredo, Alexander S Dowdell, J Scott Lee, and Sean P Colgan. 2022. "Microbial metabolite regulation of epithelial cell-cell interactions and barrier function." *Cells* 11 (6): 944.

Radzikowska, Urszula, Arturo O Rinaldi, Zeynep Çelebi Sözener, Dilara Karaguzel, Marzena Wojcik, Katarzyna Cypryk, Mübeccel Akdis, Cezmi A Akdis, and Milena Sokolowska. 2019. "The influence of dietary fatty acids on immune responses." *Nutrients* 11 (12): 2990.

Raphael, Itay, Saisha Nalawade, Todd N Eagar, and Thomas G Forsthuber. 2015. "T cell subsets and their signature cytokines in autoimmune and inflammatory diseases." *Cytokine* 74 (1): 5-17.

Rockett, Benjamin Drew, Muhammad Salameh, Kristen Carraway, Kaitlin Morrison, and Saame Raza Shaikh. 2010. "n-3 PUFA improves fatty acid composition, prevents palmitate-induced apoptosis, and differentially modifies B cell cytokine secretion in vitro and ex vivo." *Journal of lipid research* 51 (6): 1284-1297.

Rothstein, Thomas L, Daniel O Griffin, Nichol E Holodick, Tam D Quach, and Hiroaki Kaku. 2013. "Human B-1 cells take the stage." *Annals of the New York Academy of Sciences* 1285 (1): 97-114.

Saedisomeolia, Ahmad, Lisa G Wood, Manohar L Garg, Peter G Gibson, and Peter A B Wark. 2008. "Anti-inflammatory effects of long-chain n-3 PUFA in rhinovirus-infected cultured airway epithelial cells." *British journal of nutrition* 101 (4): 533-540.

Saini, Archana, Kusum Harjai, and Sanjay Chhibber. 2013. "Inhibitory effect of polyunsaturated fatty acids on apoptosis induced by Streptococcus pneumoniae in alveolar macrophages." *The Indian Journal of Medical Research* 137 (6): 1193.

Sharma, Tarang, Ashna Gupta, Ravi Chauhan, Ajaz A Bhat, Sabah Nisar, Sheema Hashem, Sabah Akhtar, Aamir Ahmad, Mohammad Haris, and Mayank Singh. 2022. "Cross-talk between the microbiome and chronic inflammation in esophageal cancer: Potential driver of oncogenesis." *Cancer and Metastasis Reviews* 41 (2): 281-299.

Shoda, Hiromi, Ryoji Yanai, Takeru Yoshimura, Tomohiko Nagai, Kazuhiro Kimura, Lucia Sobrin, Kip M Connor, Yukimi Sakoda, Koji Tamada, and Tsunehiko Ikeda. 2015. "Dietary omega-3 fatty acids suppress experimental autoimmune uveitis in association with inhibition of Th1 and Th17 cell function." *PLoS One* 10 (9): e0138241.

Snodgrass, Ryan G, and Bernhard Brüne. 2019. "Regulation and functions of 15-lipoxygenases in human macrophages." *Frontiers in pharmacology* 10: 719.

Sung, Jeehye, Heemang Jeon, In-Hwan Kim, Heon Sang Jeong, and Junsoo Lee. 2017. "Anti-inflammatory effects of stearidonic acid mediated by suppression of NF-κB and MAP-kinase pathways in macrophages." *Lipids* 52: 781-787.

Talukdar, Saswata, Eun Ju Bae, Takeshi Imamura, Hidetaka Morinaga, WuQiang Fan, Pingping Li, Wendell J Lu, Steven M Watkins, and Jerrold M Olefsky. 2010. "GPR120 is an omega-3 fatty acid receptor mediating potent anti-inflammatory and insulin-sensitizing effects." *Cell* 142 (5): 687-698.

Tarlinton, David. 2019. "B cells still front and centre in immunology." *Nature Reviews Immunology* 19 (2): 85-86.

Teague, Heather, Cassie J Fhaner, Mitchel Harris, David M Duriancik, Gavin E Reid, and Saame Raza Shaikh. 2013. "n-3 PUFAs enhance the frequency of murine B-cell subsets and restore the impairment of antibody production to a T-independent antigen in obesity [S]." *Journal of lipid research* 54 (11): 3130-3138.

Teague, Heather, Mitchel Harris, Jarrett Whelan, Sarah S Comstock, Jenifer I Fenton, and Saame Raza Shaikh. 2016. "Short-term consumption of n-3 PUFAs increases murine IL-5 levels, but IL-5 is not the mechanistic link between n-3 fatty acids and changes in B-cell populations." *The Journal of nutritional biochemistry* 28: 30-36.

Terada, Sachiyo, Mari Takizawa, Shigeru Yamamoto, Osamu Ezaki, Hiroshige Itakura, and Kiyoko S Akagawa. 2001. "Suppressive mechanisms of EPA on human T cell proliferation." *Microbiology and immunology* 45 (6): 473-481.

Tomasdottir, Valgerdur, Sigrun Thorleifsdottir, Arnor Vikingsson, Ingibjorg Hardardottir, and Jona Freysdottir. 2014. "Dietary omega-3 fatty acids enhance the B1 but not the B2 cell immune response in mice with antigen-induced peritonitis." *The Journal of nutritional biochemistry* 25 (2): 111-117.

Tull, Samantha P, Clara M Yates, Benjamin H Maskrey, Valerie B O'Donnell, Jackie Madden, Robert F Grimble, Philip C Calder, Gerard B Nash, and G Ed Rainger. 2009. "Omega-3 Fatty acids and inflammation: novel interactions reveal a new step in neutrophil recruitment." *PLoS biology* 7 (8): e1000177.

Vancolen, Seline, Guillaume Sébire, and Bernard Robaire. 2023. "Influence of androgens on the innate immune system." *Andrology*.

Verlengia, Rozangela, Renata Gorjão, Carla Cristine Kanunfre, Silvana Bordin, Thais Martins de Lima, Edgair Fernandes Martins, Philip Newsholme, and Rui Curi. 2004. "Effects of EPA and DHA on proliferation, cytokine production, and gene expression in Raji cells." *Lipids* 39: 857-864.

Wang, Hao, Qun Hao, Qiu-Rong Li, Xiao-Wen Yan, Shen Ye, You-Shen Li, Ning Li, and Jie-Shou Li. 2007. "ω-3 Polyunsaturated fatty acids affect lipopolysaccharide-induced maturation of dendritic cells through mitogen-activated protein kinases p38." *Nutrition* 23 (6): 474-482.

Wang, Ying, Jun Liu, Peter D Burrows, and Ji-Yang Wang. 2020. "B cell development and maturation." *B Cells in Immunity and Tolerance*: 1-22.

Weise, Christin, Kerstin Hilt, Milena Milovanovic, Dennis Ernst, Ralph Rühl, and Margitta Worm. 2011. "Inhibition of IgE production by docosahexaenoic acid is mediated by direct interference with STAT6 and NFκB pathway in human B cells." *The Journal of Nutritional Biochemistry* 22 (3): 269-275.

Weldon, Sinéad M, Anne C Mullen, Christine E Loscher, Lisa A Hurley, and Helen M Roche. 2007. "Docosahexaenoic acid induces an anti-inflammatory profile in lipopolysaccharide-stimulated human THP-1 macrophages more effectively than eicosapentaenoic acid." *The Journal of nutritional biochemistry* 18 (4): 250-258.

Whiting, Christine V, Paul W Bland, and John F Tarlton. 2005. "Dietary n-3 polyunsaturated fatty acids reduce disease and colonic proinflammatory cytokines in a mouse model of colitis." *Inflammatory bowel diseases* 11 (4): 340-349.

Willemsen, Linette E M, Marleen A Koetsier, Martin Balvers, Christopher Beermann, Bernd Stahl, and Eric AF van Tol. 2008. "Polyunsaturated fatty acids support epithelial barrier integrity and reduce IL-4 mediated permeability in vitro." *European journal of nutrition* 47: 183-191.

Yan, Yiqing, Wei Jiang, Thibaud Spinetti, Aubry Tardivel, Rosa Castillo, Carole Bourquin, Greta Guarda, Zhigang Tian, Jurg Tschopp, and Rongbin Zhou. 2013. "Omega-3 fatty acids prevent inflammation and metabolic disorder through inhibition of NLRP3 inflammasome activation." *Immunity* 38 (6): 1154-1163.

Yessoufou, Akadiri, Aude Plé, Kabirou Moutairou, Aziz Hichami, and Naim Akhtar Khan. 2009. "Docosahexaenoic acid reduces suppressive and migratory functions of CD4CD25 regulatory T-cells." *Journal of Lipid Research* 50 (12): 2377-2388.

Zapata-Gonzalez, Fernando, Felix Rueda, Jordi Petriz, Pere Domingo, Francesc Villarroya, Julieta Diaz-Delfin, Maria A de Madariaga, and Joan C Domingo. 2008. "Human dendritic cell activities are modulated by the omega-3 fatty acid, docosahexaenoic acid, mainly through PPARγ: RXR heterodimers: comparison with other polyunsaturated fatty acids." *Journal of Leucocyte Biology* 84 (4): 1172-1182.

Zeyda, Maximilian, Bernhard M Kirsch, René Geyeregger, Karl M Stuhlmeier, Gerhard J Zlabinger, Walter H Hörl, Marcus D Säemann, and Thomas M Stulnig. 2005. "Inhibition of human dendritic cell maturation and function by the novel immunosuppressant FK778." *Transplantation* 80 (8): 1105-1111.

Zhao, Jie, Peiliang Shi, Ye Sun, Jing Sun, Jian-Ning Dong, Hong-Gang Wang, Lu-Gen Zuo, Jian-Feng Gong, Yi Li, and Li-Li Gu. 2015. "DHA protects against experimental colitis in IL-10-deficient mice associated with the modulation of intestinal epithelial barrier function." *British Journal of Nutrition* 114 (2): 181-188.

Zurier, R B, R G Rossetti, C M Seiler, and M Laposata. 1999. "Human peripheral blood T lymphocyte proliferation after activation of the T cell receptor: effects of unsaturated fatty acids." *Prostaglandins, leukotrienes and essential fatty acids* 60 (5-6): 371-375.

Biographical Sketch

Name: *Navid Abedpoor*

Affiliation: Department of Sports Physiology, Faculty of Sports Sciences, Isfahan (Khorasgan) Branch, Islamic Azad University, Isfahan, Iran.
Department of Physiology, Medicinal Plants Research Center, Isfahan (Khorasgan) Branch, Islamic Azad University, Isfahan, Iran.

Education: Sport sciences.

Business Address: Isfahan (Khorasgan) Branch, Islamic Azad University, Isfahan, Iran.

Research and Professional Experience: Inflammation, Oxidative Stress, Non coding RNAs, Cancer, Lifestyles.

Professional Appointments: Researcher.

Honors:

1. **Won the Young Scientist** Award in "International Scientist Awards on Engineering, Science, and Medicine." 2020.
2. **Best Poster** in Royan International twin congress, Reproductive Biomedicine & Stem Cell. Protective approaches of Fraxinus

excelsior compounds on the Implantation based infertility via bioinformatics and chemoinformatic analysis.

Publications from the Last 3 Years:

1. Fatemeh Azizian-Farsani, Navid Abedpoor, Mohammad Hasan Sheikhha, Ali Osmay Gure, Mohammad Hossein Nasr Esfahani, Kamran Ghaedi. (2020). Receptor for advanced glycation end products acts as a fuel to colorectal cancer development. *Frontiers in Oncology.* IF: 6.5.
2. Fatemeh Azizian-Farsani, Marcin Osuchowski, Navid Abedpoor, Farzad Seyed Forootan, Maryam Derakhshan, Mohammad Hossein Nasr-Esfahani, Mohammad Hasan Sheikhha, Kamran Ghaedi. (2020). Anti-inflammatory and -apoptotic effects of an herbal extract on DSS-induced colitis in mice fed with high AGEs-fat diet. Scientific Reports. *Nutrition & Metabolism.* IF: 4.5.
3. Golbarg Rahimi; Salime Heydari; Bahare Rahimi; Navid Abedpoor; Iman Nicktab; Zahra Safaeinejad; Maryam Peymani; Farzad Seyed Forootan; Zahra Derakhshan; Mohammad Hossein Nasr Esfahani, Kamran Ghaedi. (2020). A combination of herbal compound (SPTC) along with exercise or metformin more efficiently alleviated diabetic complications through down-regulation of stress oxidative pathway upon activating Nrf2-Keap1 axis in AGEs rich diet-induced type 2 diabetic mice. *Nutrition and metabolism.* IF:4.5.
4. Fahimeh Akbarian, Mohsen Rahmani, Marziyeh Tavalaee, Navid Abedpoor, Mozhdeh Taki, Kamran Ghaedi, Mohammad Hossein Nasr-Esfahani. (2021). Effect of different high-fat and AGEs diets in obesity and diabetes-prone C57BL/6 mice on sperm function. *International Journal of Fertility and Sterility.* IF:2.7.
5. Navid Abedpoor, Farzaneh Taghian, Fatemeh Hajibabaie. (2022). Physical activity ameliorates the function of organs via adipose tissue in metabolic diseases. *Acta histochemical.* IF: 2.7.
6. Fatemeh Hajibabaie, Navid Abedpoor, Nazanin Asareh, Mohammad Amin Tabatabaiefar, Ali Zarrabi and Laleh Shariati. (2022). A cocktail of microRNAs as an advance diagnostic signature in stomach-colorectal cancers hallmarks incidence: a systematic review. *Personal Medicine.*
7. Navid Abedpoor, Farzaneh Taghian, Fatemeh Hajibabaie. (2022). Cross Brain-Gut Analysis Highlighted Hub Genes and LncRNAs Networks Differentially Modified During Leucine Consumption and Endurance Exercise in Mice with Depression Like Behaviors. *Molecular Neurobiology.* IF: 5.5.
8. Navid Abedpoor, Iman Niktab, Mohammad-Hossein Beigi, Masoud Baghi, Fahimeh Arzande, Naeimeh Rezaei, Mohammad-Sajad Zare, Timothy L. Megraw, Hoi-Ying Holman, Amirkianoosh Kiani, Farzad Seyed Forootan, Hossein Baharvand, Mohammad Hossein Nasr Esfahani, Kamran Ghaedi (2022). Exercise facilitates the browning of fat tissue by up-regulating Irisin receptors. (Submitted).
9. Maryam Haghparast Azad, Iman Niktab, Shaghayegh Dastjerdi, Navid Abedpoor, Golbarg Rahimi, Zahra Safaeinejad, Maryam Peymani, Farzad Seyed Forootan, Majid Asadi-Shekaari, Mohammad Hossein Nasr Esfahani & Kamran Ghaedi. The

combination of endurance exercise and SGTC (Salvia–Ginseng–Trigonella–Cinnamon) ameliorate mitochondrial markers' overexpression with sufficient ATP production in the skeletal muscle of mice fed AGEs-rich high-fat diet. (2022). *Nutrition & Metabolism.*

10. Golnaz Pakravan, Maryam Peymani, Navid Abedpoor, Zahra Safaeinejad, Mehrdad Yadegari, Maryam Derakhshan, Mohammad Hossein Nasr Esfahani, Kamran Ghaedi. Antiapoptotic and anti-inflammatory effects of Pparγ agonist, pioglitazone, reversed Dox-induced cardiotoxicity through mediating of miR-130a downregulation in C57BL/6 mice. (2022). *Journal of Biochemical and Molecular Toxicology.*

11. Fatemeh Azizian-Farsani, Navid Abedpoor, Maryam Derakhshan, Mohammad Hossein Nasr-Esfahani, Mohammad Hasan Sheikhha, Kamran Ghaedi. Protective Effects of the Combination of the Herbal Compound Against Inflammation Related to Obesity and Colitis Induced by Diet in Mice. (2022). *Iranian Journal of Diabetes and Obesity.*

12. Global multi-stakeholder endorsement of the MAFLD definition. (2022). *The Lancet Gastroenterology & Hepatology.* IF: 200.

Chapter 3

The Impacts of ALA-Based Natural Medications on Neurogenesis Statuses and Neurodegeneration Hallmarks

Navid Abedpoor[1,2]
and Fatemeh Hajibabaie[3,*]

[1]Department of Physiology, Medicinal Plants Research Center,
Isfahan (Khorasgan) Branch, Islamic Azad University, Isfahan, Iran
[2] Department of Sports Physiology, Faculty of Sports Sciences, Isfahan (Khorasgan) Branch, Islamic Azad University, Isfahan, Iran
[2]Department of Biology, Faculty of Basic Sciences, Shahrekord Branch, Islamic Azad University, Shahrekord, Iran

Abstract

The maintenance of neurogenesis is a critical aspect of brain restoration after injury. Neurogenesis is affected by various intrinsic factors, such as oxidative stress, inflammation, aging, and extrinsic factors, like environmental pollution, lifestyle, and diet. Considering the increasing proportion of the elderly worldwide, searching for new therapeutic targets to tackle brain aging and related neurodegenerative diseases has become one of the most pressing and challenging in present-day biomedicine.

A wide range of neurological, psychiatric, and developmental brain disorders has been linked to the dysregulation of fatty acid and phospholipid metabolism in the central nervous system (CNS). Alpha-linolenic acid, an essential polyunsaturated fatty acid with antioxidant, anti-inflammatory, and antiapoptotic properties, is highly concentrated in

---

[*] Corresponding Author's Email: fateme.hajibabaii1991@gmail.com.

In: Properties and Uses of Linolenic Acid
Editor: Calvin S. Willmon
ISBN: 979-8-89113-201-6
© 2023 Nova Science Publishers, Inc.

the brain. Considering the evident empirical evidence that showcases the positive features of plant-based nutrition and long-chain n-3 fatty acids, there is considerable interest in investigating the effects of the plant-based n-3 fatty acid alpha-linolenic acid (ALA) on neurogenesis and brain injuries. This chapter briefly overviews the numerous neurogenesis attractive features of ALA and reviews the existing literature on the subject.

Introduction

Neurogenesis maintenance is crucial for repairing the brain after injury. It is believed that neurogenesis has a role in stress responses and higher functions like brain plasticity, which includes cognition, memory, mood, and perceptual learning (olfactory) (Hajibabaie et al., 2023). The development of seizures, depression, and a decrease in learning abilities have been linked to a reduction in neurogenesis, as observed during aging or in pathological conditions. Reducing the number and/or function of neural stem cells (NSCs) and neural progenitor cells (NPCs) could lead to impaired neurogenesis. This phenomenon could be attributed to the combined efforts of multiple mechanisms that operate in the brain during aging or neurodegenerative conditions, including inflammation and oxidative stress. Toxic substances such as short-chain fatty acids (SCFAs), branched-chain amino acids, and peptidoglycans produced by an altered intestinal microbiota might impair neurogenesis (Abedpoor, Taghian, and Hajibabaie 2022). The microbial communities in the gut are influenced and regulated by various external factors, including lifestyle and dietary. It is important to understand that imbalances within this complex ecosystem significantly impact the body's various barriers, including the blood-brain barrier (BBB) and the enteric barrier. These imbalances can compromise the permeability of these barriers, resulting in the passage of harmful substances along the gut-brain axis (GBA) and into brain tissue (Abedpoor, Taghian, and Hajibabaie 2022). The exogenous regulation of gut microbiota composition mechanisms is interesting in countering neurogenesis deterioration. The gut microbiota consists of multiple microorganisms living together in a fragile equilibrium, such as bacteria, yeast, and viruses. An imbalance in this complex, known as dysbiosis, can result in atypical neural and glial reactivity as well as the loss of neurogenic capacity (Abedpoor, Taghian, and Hajibabaie 2022).

Consequently, there exists a functional correlation between microbiota, GBA, and neurogenesis, with imbalances in this axis affecting not just the neural control of the gastrointestinal tract, but also instigating various brain disorders such as mood disorders (e.g., depression, and anxiety), neurodevelopmental disorders (e.g., autism), and cognitive disorders (e.g., Alzheimer's disease).

GBA exploits several anatomic structures, systems, and metabolic routes to establish a bidirectional connection between enteric microbes and the brain (Abedpoor, Taghian, and Hajibabaie 2022). These include the neuroendocrine system (by the hypothalamic pituitary–adrenal (HPA) axis), the neuro-immune system, as well as the sympathetic and parasympathetic arms of the autonomic nervous system, including the enteric nervous system, the vagus nerve, and the immune system. Unsurprisingly, the GBA has been characterized as a "second brain" (Chan et al., 2022; Toledo et al., 2022; Sorboni et al., 2022).

The composition of microbiota can be influenced by numerous factors, such as infection, mode of birth delivery, use of antibiotic medications, the nature of the nutritional provision, environmental stressors, host genetics, and aging, which increase the level of complexity. Due to its practicality, modifying diet composition is an attractive therapeutic approach for targeting GBA and neurogenesis through the microbiota (Sarubbo, Cavallucci, and Pani 2022). The possible adjuvant effects of natural antioxidants and anti-inflammatory molecules, like dietary polyphenols, on adult neurogenesis (AN) have been thoroughly studied (Varesi et al., 2022).

In brief, preserving a healthy brain in the lifespan may require balancing the gut microbiota with considering correct diet and sustaining GBA.

Adult hippocampal neurogenesis is fundamental for learning, memory, and mood regulation; however, it tends to decline with age and illness. The exact mechanism through which the orphan nuclear receptor TLX/NR2E1 regulates neural stem and progenitor cell self-renewal and proliferation is yet to be determined (Kandel et al., 2020). According to studies, the survival and proliferation of neural stem and progenitor cells are contingent on the presence of mono-unsaturated fatty acids (Dahlmann 2019).

It has been suggested that adult neurogenesis may play a crucial role in repairing the brain after damage has occurred, as well as participating in processes related to plasticity, such as memory, cognition, mood, and sensory functions (Sarubbo, Cavallucci, and Pani 2022; Chen, Epstein, and Stern 2010). Numerous intrinsic factors are believed to affect neurogenesis, including oxidative stress, inflammation, aging, and extrinsic factors such as environmental pollution, lifestyle, and diet (Abedpoor, Taghian, and Hajibabaie 2022). However, there has been increasing focus on the impact of the gut microbiota, which comprises many saprophytic microorganismal communities inhabiting the intestinal ecosystem (Gebrayel et al., 2022). There is an increasing amount of evidence that comes mainly from animal studies that demonstrate how the microbiota and its imbalances, which are associated with diseases, affect the activities of neural stem cell proliferation and differentiation in the neurogenic niches of the brain (Gao et al., 2021). Conversely, it is possible that the alleged pro-neurogenic function of natural dietary substances containing antioxidants and anti-inflammatory properties, such as polyphenols, polyunsaturated fatty acids, and pro/prebiotics, may be influenced, to some extent, by their impact on the microflora of the intestine (Sarubbo, Cavallucci, and Pani 2022; Romanenko et al., 2021). This chapter aims to summarize the currently available information regarding the influence of alpha-linolenic acid (ALA) on neurogenesis, as well as to analyze the potential underlying mechanisms that may be involved in this process. Additionally, we will discuss the potential implications of this emerging knowledge for the fight against neurodegeneration and brain aging, which could be of great importance in the field of neuroscience.

Given the increasing proportion of the elderly population worldwide, searching for new therapeutic targets against brain aging and associated neurodegenerative diseases is one of the most urgent and challenging issues in current biomedicine (Hajibabaie et al., 2023). The causes of aging and neurodegeneration are numerous, and some of the most prominent ones include the limited renewal capacity of brain cells, the alteration of brain vasculature, and neuronal/glial dysfunction (Hajibabaie et al., 2023; Abedpoor, Taghian, and Hajibabaie 2022). With aging, the brain damage repair systems, which include adult neurogenesis, also experience a decline. The process of neurogenesis involves the generation of new neurons, glial cells, and other neural lineages, which are derived from neural stem cells (NSCs) and neural progenitor cells (NPCs) (Gage and Temple 2013). An important part of this process is the integration of NSCs or NPSs into the preexisting neuronal network, which can only occur after maturation, migration, and functional integration. The term adult neurogenesis (AN) is used to describe the phenomenon when it takes place in adult life (de Miranda et al., 2017). While NSCs can be found in various brain regions, primarily the hippocampus's sub-granular zone and the lateral ventricle subventricular zone serve as the main AN niches. Although adult neurogenesis in the neocortex, striatum, amygdala, and substantia nigra is restricted under normal physiological conditions, it is possible to stimulate it following an injury.

The Role of Linolenic Acid in Neurogenesis Maintenance

The category of unsaturated fatty acids that are essential and contain more than one double bond (C = C) is known as polyunsaturated fatty acids (PUFAs) (Hajibabaie et al., 2022). These crucial nutrients must primarily be acquired through diet or supplements, as mammals cannot support their de novo synthesis. The two primary biologically important categories of long-chain PUFAs are n-6 PUFAs (omega-6) and n-3 PUFAs (omega-3), with their first double bond located at C6 or C3, respectively, counted from the methyl C (Hajibabaie et al., 2022). Moreover, it is worth noting that Linoleic acid (n-6 PUFA) and α-linolenic acid (n-3 PUFA) are considered essential fatty acids because of their inability to be synthesized by the human body and their role as the precursors of other critical PUFAs (Michalak, Mosińska, and Fichna 2016; Iizuka 2021). Alpha-linolenic acid, eicosapentaenoic acid, and docosahexaenoic acid are nutritionally important PUFAs with antioxidant, anti-inflammatory, and antiapoptotic properties that have high concentrations in the brain (Crupi, Marino, and Cuzzocrea 2013). Based on previous evidence, the intestinal microbial ecosystem is notably impacted by the presence of dietary PUFAs (Marrone and Coccurello 2019).

On the other hand, the overproduction of n-6 PUFAs and increased n-6 to n-3 ratio in mice was associated with systemic inflammation, as well as intestinal dysbiosis characterized by an increase in Proteobacteria and a reduction in Bacteroides and Actinobacteria, and abnormal gut permeability. In a similar study, however, the transgenic amplification of n-3 PUFAs tissue content inhibited inflammation induced by LPS while also promoting the growth of Bifidobacterium and maintaining the integrity of the intestinal barrier (Kaliannan et al., 2015).

The authors focused on metabolic endotoxemia as the principal dysbiosis-associated disorder in transgenic animals. Nevertheless, research highlights a potential PUFA microbiota-brain signaling pathway that requires further examination. The immature precursors of the central nervous system (CNS) are known as embryonic neural stem cells (eNSCs), which possess the capacity for self-renewal and multipotential differentiation. Alpha-linolenic acid (ALA), a plant-based essential omega-3 polyunsaturated fatty acid, is among the endogenous and exogenous factors that regulate eNSCs (Barmak et al.,).

Eicosapentaenoic acid (EPA) and Docosahexaenoic acid (DHA), both members of the Omega-3 family of PUFAs, are necessary for the development of the brain and comprise about 8% of its weight (O'Brien and Sampson 1965). Both EPA and DHA are pivotal in the structure and function of neurons (Robinson, Ijioma, and Harris 2010; Eckert et al., 2010). The current focus is on investigating the importance and role of other fatty acids (FAs). According to recent findings, alpha-linolenic acid (ALA), an essential omega-3 fatty acid, is responsible for the increase in the proliferation and differentiation of cultured neural stem cells (NSCs) (Hejr et al., 2017). The proliferative activity of β-sitosterol on the NSCs *in vitro* conditions was previously reported (Jiang et al., 2013). Nevertheless, certain studies have indicated the differentiating activity of stearic acid (SA) and myristic acid (MA) in the brain, two frequently occurring saturated FAs with 18 and 14 carbons, respectively (Gozlan-Devillierre, Baumann, and Bourre 1976; Rioux et al., 2008). Among the diverse sources of herbal fatty acids, Alyssum homolocarpum seed oil (AHSO)-rich in natural alpha-linolenic acid, stearic acid, myristic acid and β-sitosterol- is marked for its high concentration and its ability to induce NSC proliferation without affecting NSC differentiation (Hamedi et al., 2015). Alyssum, a well-known genus within the Brassicaceae family, commonly referred to as the mustard family, is native to the Middle East, specifically Iran, Iraq, and Pakistan, and encompasses approximately 100-170 related species (Mahmoudi et al., 2019).

Susana García-Cerro and co-workers indicated that consumption of linolenic acid enhanced the brain weight and had beneficial cognitive and neuromorphological effects in down syndrome mice (García-Cerro et al., 2020; Avila-Martin et al., 2011). Growing studies have demonstrated that linolenic acid could cross the blood-brain barrier and placental. Notably, in down syndrome, the concentration of MUFAs significantly decreased in the brain (Avila-Martin et al., 2011; Bloch and Qawasmi 2011). Therefore, the lower concentration of the MUFAs, especially linolenic acid, might be essential in neurodevelopmental (Bloch and Qawasmi 2011; Liu et al., 2003).

It has been established that administering linolenic acid could improve the development of brain and neurogenesis during the first stages of postnatal life and gestation. Based on these data, ω-3 fatty acids is required to promote hippocampal neurogenesis and conversion into Docosahexaenoic acid (DHA) (Liu et al., 2003; Cao et al., 2009; Niculescu, Lupu, and Craciunescu 2011).

Additionally, the linolenic acid treatment elevated the quantity of fully developed granular cells, conceivably owing to the acknowledged impact of both acids on neuronal differentiation and survival (Niculescu, Lupu, and Craciunescu 2011; Blondeau et al., 2009). The augmented cellularity is probably the cause for the amplified granular cell layer (GCL) volume and brain weight observed in down syndrome mice following prenatal treatment with linolenic acid. The administration of linolenic acid has been demonstrated by studies to result in an increase in hippocampal volume (Blondeau et al., 2009; Cutuli et al., 2016).

The administration of exogenous fatty acids to pregnant down syndrome mice may have a positive impact on prenatal brain development, restoring the size of the hippocampal structures in their offspring due to the low concentrations of fatty acids in down syndrome brains (Shah 1979; Zmijewski et al., 2015).

Nevertheless, the majority of advantageous impacts of the antenatal delivery of linolenic acids on the neuromorphological changes observed in down syndrome mice were not sustained six weeks following the cessation of the intervention. The findings suggest that linolenic acid are required to stimulate their pro-survival and proliferative effects in down syndrome mice (Zmijewski et al., 2015).

Administration of fatty acids during adulthood also reduces the neuropathology in down syndrome mice (García-Cerro et al., 2020). One study by Giacomini et al., demonstrated that administering corn oil, which comprises oleic and linoleic acids, restored brain weight, neurogenesis, dendritogenesis, and cognition in adult down syndrome mice (Giacomini et al., 2018). Furthermore, the authors showed that NPCs derived from DS individuals can increase proliferation rates when treated with linolenic and oleic acids (Giacomini et al., 2018).

Adult neurogenesis has been observed in certain areas of the brain, such as the hypothalamus, subventricular zone, and subgranular zone in the hippocampi. It has been found through recent studies that diet plays a role in the development of new cells in the hypothalamus. Safflower seed oil (SSO) has been found to have beneficial effects, as demonstrated by research studies, and it is known for being a rich source of linoleic acid (Giacomini et al., 2018). Mehrzad Jafari Barmak and colleagues have revealed the impact of safflower seed oil on neurogenesis compared to synthetic linoleic acid. In this study, the adult mice consumed with safflower seed oil (400 mg/kg) and pure synthetic linoleic acid (300 mg/kg) (Barmak et al.,). Mehrzad Jafari Barmak has found that neurogenesis and proliferation improved in the hypothalamic (Barmak et al.,). Moreover, consuming Safflower seed oil enriched with linoleic acid enhances the serum brain-derived neurotrophic factor (BNDF).

The hypothalamic nuclei play a crucial role in regulating food intake and related body functions, and it has been found that neurogenesis and the survival of neurons are essential factors in this process. Aging, chronic stress, and central nervous system disorders significantly reduce neural stem/progenitor cell proliferation and self-renewal. According to research, a possible outcome of this is weight gain and related diseases (Calapai 2008). In developing countries, herbal therapy is the preferred treatment option for most diseases, and it is the first line of treatment, as noted in (Kamelia et al., 2016). Recently, evidence indicated that linoleic acid consumption could benefit the proliferation and differentiation in the hypothalamic (Kamelia et al., 2016).

In neuroscience, inducing neurogenesis through dietary factors is seen as a promising therapeutic approach, according to reference (Poulose et al., 2017). Altering exogenous factors like diet is more practical and easier to stimulate neurogenesis than manipulating endogenous factors like genetic networks. Proper neural development and function depend on critical nutritional factors, among which PUFAs have been identified (Santos et al., 2008; Barmak et al.,). For a long time, it has been known that PUFAs are not synthesized in vertebrates and are produced in plants (Kelly 1984). The n-6 PUFA, Linoleic acid (LA, C18:2n-6), is predominantly found in plant oils like soybean and corn, commonly used in western diets (Barnes, Bloom, and Nahin 2008). The results of the study revealed that Safflower, which is known to be a rich source of linoleic acid (73.64%), had a significant impact on the proliferative and differentiative capacities of neural stem cells (NSCs), leading to an increase in the number of neurons (β-III tubulin positive cells). In contrast, synthetic LA, at similar concentrations to natural SSO LA, could not affect the number of neurons (Ghareghani et al., 2017).

According to research, several intrinsic factors have been shown to stimulate hypothalamic neurogenesis. These factors include fibroblast growth factor 2, insulin-like growth factor (Pérez-Martín et al., 2010), and BNDF (Pencea et al., 2001). One of the reasons why BDNF is an important factor in hypothalamic neurogenesis is that it has a high potential to transit the blood-brain barrier (BBB) in both directions, as demonstrated by research (Pan et al., 1998). In addition, according to previous research, the levels of BDNF in the serum have been suggested to serve as a significant reserve pool for the brain (Laske et al., 2011). The administration of SSO or LA was found to have a significant effect on BDNF serum levels. One important point to note is that SSO had a significantly more potent effect on BDNF serum levels than LA.

Since BDNF has the potential to cross the blood-brain barrier (BBB), measuring the levels of serum, BDNF could serve as a useful indicator of its levels in the brain (Laske et al., 2011).

In another study, Pidoplichko and co-workers have demonstrated that the consumption of linolenic acid might increase the simplification of GABAergic neurotransmission in the hippocampus of the 6- to 10-week-old male, Sprague-Dawley rats (Pidoplichko et al., 2023).

The reduction of neuronal cell death can be attributed to linolenic acid, which achieves this by inhibiting glutamatergic neurotransmission. This inhibition is brought about by activating a relatively new class of potassium channels, known as TREK-1, which has two-pore-forming domains. TREK-1 channels are known to have potassium background channel function, are abundantly expressed in the brain, and are localized on both presynaptic and postsynaptic neurons (Fink et al., 1996; Patel et al., 1998; Kim et al., 1995; Lauritzen et al., 2000).

The activation of TREK-1 potassium channels by linolenic acid results in the hyperpolarization of synaptic terminal neuronal membranes, leading to decreased glutamate release and hyperpolarization of postsynaptic neuronal membranes. This, in turn, enables the blockade of the magnesium-associated channel of the N-methyl-D-aspartate (NMDA) receptor and inhibits the depolarization caused by the action of glutamate on other glutamate ionotropic receptors(Lauritzen et al., 2000).

Pidoplichko and colleagues revealed that 24 h after a single subcutaneous injection of linolenic acid enhanced $GABA_AR$-mediated suppressor in hippocampus injection in 6- to 10-week-old male, Sprague-Dawley rats (Pidoplichko et al., 2023; Lauritzen et al., 2000).

As a result of these findings, we were compelled to conduct further studies on this effect. For this purpose, we applied linolenic acid (100 µM) to slices obtained from young adult male rats. The GABAAR inhibitory activity was observed to have increased following administration of ALA, as evidenced by an increase in the charge transferred by outward GABAergic currents in voltage-clamp mode (Pidoplichko et al., 2023; Lauritzen et al., 2000). Linolenic acid has been observed to facilitate inhibitory activity in the hippocampal region, which has an effect (Pidoplichko et al., 2023). The reversible nature of the effect was overshadowed by the significant washout time, indicating that the treatment had a prolonged impact (Pidoplichko et al., 2023).

Conclusion

Despite a considerable amount of evidence indicating the significant role played by intestinal microbiota and bacteria-derived metabolites in the gut-brain communication axis, especially in the modulation of AN, the consolidation of this crucial information and its potential application to clinical practice are still confronted with several formidable challenges. To accurately identify compounds potentially active on neurogenesis and gut microbiota, it is necessary to conduct more in-depth fundamental research on animal and cell culture models, considering their dosing, bio assimilation, and synergistic effect, both in physiological and pathological situations. Further, a clinical investigation is necessary to assess the effectiveness of potential molecules, notably natural plant-based antioxidants and anti-inflammatory compounds, regarding their neuroprotective and potentially neurogenic properties.

References

Abedpoor, N., Farzaneh Taghian, and Fatemeh Hajibabaie. 2022. "Cross brain-gut analysis highlighted hub genes and LncRNA networks differentially modified during leucine consumption and endurance exercise in mice with depression-like behaviors." *Molecular Neurobiology* 59 (7): 4106-4123.

Avila-Martin, G., Iriana Galan-Arriero, Julio Gomez-Soriano, and Julian Taylor. 2011. "Treatment of rat spinal cord injury with the neurotrophic factor albumin-oleic acid: translational application for paralysis, spasticity and pain." *PLoS One* 6 (10): e26107.

Barmak, M. J., Ebrahim Nouri, Maryam Hashemi Shahraki, Ghasem Ghalamfarsa, Kazem Zibara, Hamdallah Delaviz, and Amir Ghanbari. "Safflower seed oil, a rich source of linoleic acid, stimulates hypothalamic neurogenesis *in vivo*." *Anatomy & cell biology*.

Barnes, P. M., Barbara Bloom, and Richard L. Nahin. 2008. "Complementary and alternative medicine use among adults and children: United States, 2007."

Bloch, M. H., and Ahmad Qawasmi. 2011. "Omega-3 fatty acid supplementation for the treatment of children with attention-deficit/hyperactivity disorder symptomatology: systematic review and meta-analysis." *Journal of the American Academy of Child & Adolescent Psychiatry* 50 (10): 991-1000.

Blondeau, N., Carine Nguemeni, David N. Debruyne, Marie Piens, Xuan Wu, Hongna Pan, XianZhang Hu, Carine Gandin, Robert H Lipsky, and Jean-Christophe Plumier. 2009. "Subchronic alpha-linolenic acid treatment enhances brain plasticity and exerts an antidepressant effect: a versatile potential therapy for stroke." *Neuropsychopharmacology* 34 (12): 2548-2559.

Calapai, G. 2008. "European legislation on herbal medicines: a look into the future." *Drug Safety* 31: 428-431.

Cao, D., Karl Kevala, Jeffrey Kim, Hyun-Seuk Moon, Sang Beom Jun, David Lovinger, and Hee-Yong Kim. 2009. "Docosahexaenoic acid promotes hippocampal neuronal development and synaptic function." *Journal of neurochemistry* 111 (2): 510-521.

Chan, D. G., Katelyn Ventura, Ally Villeneuve, Paul Du Bois, and Matthew R Holahan. 2022. "Exploring the Connection Between the Gut Microbiome and Parkinson's Disease Symptom Progression and Pathology: Implications for Supplementary Treatment Options." *Journal of Parkinson's Disease* 12 (8): 2339-2352.

Chen, H., Jane Epstein, and Emily Stern. 2010. "Neural plasticity after acquired brain injury: evidence from functional neuroimaging." *PM&R* 2: S306-S312.

Crupi, R., Angela Marino, and Salvatore Cuzzocrea. 2013. "n-3 fatty acids: role in neurogenesis and neuroplasticity." *Current medicinal chemistry* 20 (24): 2953-2963.

Cutuli, D., Marco Pagani, Paola Caporali, Alberto Galbusera, Daniela Laricchiuta, Francesca Foti, Cristina Neri, Gianfranco Spalletta, Carlo Caltagirone, and Laura Petrosini. 2016. "Effects of omega-3 fatty acid supplementation on cognitive functions and neural substrates: a voxel-based morphometry study in aged mice." *Frontiers in aging neuroscience* 8: 38.

Dahlmann, E. 2019. "HCMV manipulation of host cholesteryl ester metabolism." The University of Arizona.

de Miranda, A. S., Cun-Jin Zhang, Atsuko Katsumoto, and Antônio Lúcio Teixeira. 2017. "Hippocampal adult neurogenesis: Does the immune system matter?" *Journal of the Neurological Sciences* 372: 482-495.

Eckert, Gunter P, Cornelia Franke, Michael Nöldner, Oliver Rau, Mario Wurglics, Manfred Schubert-Zsilavecz, and Walter E Müller. 2010. "Plant derived omega-3-fatty acids protect mitochondrial function in the brain." *Pharmacological research* 61 (3): 234-241.

Fink, M., Fabrice Duprat, Florian Lesage, Roberto Reyes, Georges Romey, Catherine Heurteaux, and M Lazdunski. 1996. "Cloning, functional expression and brain localization of a novel unconventional outward rectifier K+ channel." *The EMBO journal* 15 (24): 6854-6862.

Gage, F. H., and Sally Temple. 2013. "Neural stem cells: generating and regenerating the brain." *Neuron* 80 (3): 588-601.

Gao, J., Yuan Liao, Mengsheng Qiu, and Wanhua Shen. 2021. "Wnt/β-catenin signaling in neural stem cell homeostasis and neurological diseases." *The Neuroscientist* 27 (1): 58-72.

García-Cerro, S., Noemí Rueda, Verónica Vidal, Alba Puente, Víctor Campa, Sara Lantigua, Oriol Narcís, Ana Velasco, Renata Bartesaghi, and Carmen Martínez-Cué. 2020. "Prenatal administration of oleic acid or linolenic acid reduces neuromorphological and cognitive alterations in Ts65dn down syndrome mice." *The Journal of nutrition* 150 (6): 1631-1643.

Gebrayel, P., Carole Nicco, Souhaila Al Khodor, Jaroslaw Bilinski, Elisabetta Caselli, Elena M Comelli, Markus Egert, Cristina Giaroni, Tomasz M Karpinski, and Igor Loniewski. 2022. "Microbiota medicine: Towards clinical revolution." *Journal of Translational Medicine* 20 (1): 1-20.

Ghareghani, M., Kazem Zibara, Hassan Azari, Hossein Hejr, Farzad Sadri, Ramin Jannesar, Ghasem Ghalamfarsa, Hamdallah Delaviz, Ebrahim Nouri, and Amir Ghanbari. 2017.

"Safflower seed oil, containing oleic acid and palmitic acid, enhances the stemness of cultured embryonic neural stem cells through Notch1 and induces neuronal differentiation." *Frontiers in Neuroscience* 11: 446.

Giacomini, A., Fiorenza Stagni, Marco Emili, Sandra Guidi, Maria Elisa Salvalai, Mariagrazia Grilli, Veronica Vidal-Sanchez, Carmen Martinez-Cué, and Renata Bartesaghi. 2018. "Treatment with corn oil improves neurogenesis and cognitive performance in the Ts65Dn mouse model of Down syndrome." *Brain research bulletin* 140: 378-391.

Gozlan-Devillierre, N., N. Baumann, and J. M. Bourre. 1976. "Study of the passage of stearic acid through blood-brain barrier and its incorporation in cerebral membranes (especially in myelin)." *Comptes Rendus Hebdomadaires des Seances de L'academie des sciences. Serie D: Sciences Naturelles* 282 (20): 1825-1828.

Hajibabaie, F., Navid Abedpoor, Farzaneh Taghian, and Kamran Safavi. 2023. "A cocktail of polyherbal bioactive compounds and regular mobility training as senolytic approaches in age-dependent alzheimer's: the in silico analysis, lifestyle intervention in old age." *Journal of Molecular Neuroscience* 73 (2-3): 171-184.

Hajibabaie, F., Navid Abedpoor, Kamran Safavi, and Farzaneh Taghian. 2022. "Natural remedies medicine derived from flaxseed (secoisolariciresinol diglucoside, lignans, and α-linolenic acid) improve network targeting efficiency of diabetic heart conditions based on computational chemistry techniques and pharmacophore modeling." *Journal of Food Biochemistry* 46 (12): e14480.

Hamedi, A., Amir Ghanbari, Razieh Razavipour, Vahid Saeidi, Mohammad M Zarshenas, Maryam Sohrabpour, and Hassan Azari. 2015. "Alyssum homolocarpum seeds: phytochemical analysis and effects of the seed oil on neural stem cell proliferation and differentiation." *Journal of natural medicines* 69: 387-396.

Hejr, H., Majid Gharegham, Kazem Zibara, Maryam Ghafari, Farzad Sadri, Zinab Salehpour, Azadeh Hamedi, Koresh Negintaji, Hassan Azari, and Amir Ghanbari. 2017. "The ratio of 1/3 linoleic acid to alpha linolenic acid is optimal for oligodendrogenesis of embryonic neural stem cells." *Neuroscience letters* 651: 216-225.

Iizuka, K. 2021. "The roles of carbohydrate response element binding protein in the relationship between carbohydrate intake and diseases." *International journal of molecular sciences* 22 (21): 12058.

Jiang, L., Nian-yun Yang, Xiao-lin Yuan, Yi-jie Zou, Ze-qun Jiang, Feng-ming Zhao, Jianping Chen, Ming-yan Wang, and Da-xiang Lu. 2013. "Microarray analysis of mRNA and microRNA expression profile reveals the role of β-sitosterol-D-glucoside in the proliferation of neural stem cell." *Evidence-Based Complementary and Alternative Medicine* 2013.

Kaliannan, K., Bin Wang, Xiang-Yong Li, Kui-Jin Kim, and Jing X. Kang. 2015. "A hostmicrobiome interaction mediates the opposing effects of omega-6 and omega-3 fatty acids on metabolic endotoxemia." *Scientific reports* 5 (1): 11276.

Kamelia, E., Hadiyat Miko, Marni Br Karo, and Mochammad Hatta. 2016. "Neurogenesis and brain-derived neurotrophic factor levels in herbal therapy." *International Journal of Research in Medical Sciences* 4 (11): 4654.

Kandel, P., Fatih Semerci, Aleksandar Bajic, Dodge Baluya, LiHua Ma, Kevin Chen, Austin Cao, Tipwarin Phongmekhin, Nick Matinyan, and William Choi. 2020. "Oleic acid triggers hippocampal neurogenesis by binding to TLX/NR2E1." *bioRxiv*: 2020.10. 28.359810.

Kelly, G. J. 1984. "Formation and fates of plant polyunsaturated fatty acids."

Kim, D., Celia D Sladek, Carmen Aguado-Velasco, and Joanne R Mathiasen. 1995. "Arachidonic acid activation of a new family of K+ channels in cultured rat neuronal cells." *The Journal of Physiology* 484 (3): 643-660.

Laske, C., Konstantinos Stellos, Nadine Hoffmann, Elke Stransky, Guido Straten, Gerhard W Eschweiler, and Thomas Leyhe. 2011. "Higher BDNF serum levels predict slower cognitive decline in Alzheimer's disease patients." *International Journal of Neuropsychopharmacology* 14 (3): 399-404.

Lauritzen, I., Nicolas Blondeau, Catherine Heurteaux, Catherine Widmann, Georges Romey, and Michel Lazdunski. 2000. "Polyunsaturated fatty acids are potent neuroprotectors." *The EMBO journal* 19 (8): 1784-1793.

Liu, D. P., Cecilia Schmidt, Timothy Billings, and Muriel T. Davisson. 2003. "Quantitative PCR genotyping assay for the Ts65Dn mouse model of Down syndrome." *Biotechniques* 35 (6): 1170-1180.

Mahmoudi, R., Majid Ghareghani, Kazem Zibara, Maryam Tajali Ardakani, Yahya Jand, Hassan Azari, Jafar Nikbakht, and Amir Ghanbari. 2019. "Alyssum homolocarpum seed oil (AHSO), containing natural alpha linolenic acid, stearic acid, myristic acid and β-sitosterol, increases proliferation and differentiation of neural stem cells *in vitro*." *BMC Complementary and Alternative Medicine* 19 (1): 1-11.

Marrone, M. C., and Roberto Coccurello. 2019. "Dietary fatty acids and microbiota-brain communication in neuropsychiatric diseases." *Biomolecules* 10 (1): 12.

Michalak, A., Paula Mosińska, and Jakub Fichna. 2016. "Polyunsaturated fatty acids and their derivatives: therapeutic value for inflammatory, functional gastrointestinal disorders, and colorectal cancer." *Frontiers in pharmacology* 7: 459.

Niculescu, M. D., Daniel S. Lupu, and Corneliu N. Craciunescu. 2011. "Maternal α-linolenic acid availability during gestation and lactation alters the postnatal hippocampal development in the mouse offspring." *International Journal of Developmental Neuroscience* 29 (8): 795-802.

O'Brien, J. S., and E. Lois Sampson. 1965. "Lipid composition of the normal human brain: gray matter, white matter, and myelin." *Journal of lipid research* 6 (4): 537-544.

Pan, W., William A Banks, Melita B. Fasold, Jonathan Bluth, and Abba J. Kastin. 1998. "Transport of brain-derived neurotrophic factor across the blood–brain barrier." *Neuropharmacology* 37 (12): 1553-1561.

Patel, A. J., Eric Honoré, François Maingret, Florian Lesage, Michel Fink, Fabrice Duprat, and Michel Lazdunski. 1998. "A mammalian two pore domain mechano-gated S-like K+ channel." *The EMBO journal* 17 (15): 4283-4290.

Pencea, V., Kimberly D. Bingaman, Stanley J. Wiegand, and Marla B. Luskin. 2001. "Infusion of brain-derived neurotrophic factor into the lateral ventricle of the adult rat leads to new neurons in the parenchyma of the striatum, septum, thalamus, and hypothalamus." *Journal of neuroscience* 21 (17): 6706-6717.

Pérez-Martín, M., M. Cifuentes, J. M. Grondona, M. D. López-Avalos, U. Gómez-Pinedo, J. M. García-Verdugo, and P. Fernández-Llebrez. 2010. "IGF-I stimulates neurogenesis in the hypothalamus of adult rats." *European Journal of Neuroscience* 31 (9): 1533-1548.

Pidoplichko, V. I., Taiza H. Figueiredo, Maria F. M. Braga, Hongna Pan, and Ann M. Marini. 2023. "Alpha-linolenic acid enhances the facilitation of GABAergic neurotransmission in the BLA and CA1." *Experimental Biology and Medicine*: 15353702231165010.

Poulose, S. M., Marshall G. Miller, Tammy Scott, and Barbara Shukitt-Hale. 2017. "Nutritional factors affecting adult neurogenesis and cognitive function." *Advances in nutrition* 8 (6): 804-811.

Rioux, V., D. Catheline, E. Beauchamp, J. Le Bloc'h, F. Pédrono, and P. Legrand. 2008. "Substitution of dietary oleic acid for myristic acid increases the tissue storage of α-linolenic acid and the concentration of docosahexaenoic acid in the brain, red blood cells and plasma in the rat." *Animal* 2 (4): 636-644.

Robinson, J. G., Nkechinyere Ijioma, and William Harris. 2010. "Omega-3 fatty acids and cognitive function in women." *Women's Health* 6 (1): 119-134.

Romanenko, M., Victor Kholin, Alexander Koliada, and Alexander Vaiserman. 2021. "Nutrition, gut microbiota, and Alzheimer's disease." *Frontiers in psychiatry* 12: 712673.

Santos, L. M. dos, Darci Neves dos Santos, Ana Cecília Sousa Bastos, Ana Marlúcia Oliveira Assis, Matildes Silva Prado, and Mauricio L Barreto. 2008. "Determinants of early cognitive development: hierarchical analysis of a longitudinal study." *Cadernos de Saúde Pública* 24: 427-437.

Sarubbo, F., Virve Cavallucci, and Giovambattista Pani. 2022. "The influence of gut microbiota on neurogenesis: evidence and hopes." *Cells* 11 (3): 382.

Shah, S. N. 1979. "Fatty acid composition of lipids of human brain myelin and synaptosomes: changes in phenylketonuria and Down's syndrome." *International Journal of Biochemistry* 10 (6): 477-482.

Sorboni, S. G., Hanieh Shakeri Moghaddam, Reza Jafarzadeh-Esfehani, and Saman Soleimanpour. 2022. "A comprehensive review on the role of the gut microbiome in human neurological disorders." *Clinical Microbiology Reviews* 35 (1): e00338-20.

Toledo, A. R. L., Germán Rivera Monroy, Felipe Esparza Salazar, Jea-Young Lee, Shalini Jain, Hariom Yadav, and Cesario Venturina Borlongan. 2022. "Gut–brain axis as a pathological and therapeutic target for neurodegenerative disorders." *International Journal of Molecular Sciences* 23 (3): 1184.

Varesi, A., Elisa Pierella, Marcello Romeo, Gaia Bavestrello Piccini, Claudia Alfano, Geir Bjørklund, Abigail Oppong, Giovanni Ricevuti, Ciro Esposito, and Salvatore Chirumbolo. 2022. "The potential role of gut microbiota in Alzheimer's disease: From diagnosis to treatment." *Nutrients* 14 (3): 668.

Zmijewski, P. A., Linda Y. Gao, Abhinav R. Saxena, Nastacia K. Chavannes, Shazaan F. Hushmendy, Devang L. Bhoiwala, and Dana R. Crawford. 2015. "Fish oil improves gene targets of Down syndrome in C57BL and BALB/c mice." *Nutrition research* 35 (5): 440-448.

Biographical Sketch

Name: *Fatemeh Hajibabaie*

Affiliation:
Department of Biology, Faculty of Basic Sciences, Shahrekord Branch, Islamic Azad University, Shahrekord, Iran.
Department of Physiology, Medicinal Plants Research Center, Isfahan (Khorasgan) Branch, Islamic Azad University, Isfahan, Iran

Education: Molecular genetics.

Business Address: Isfahan (Khorasgan) Branch, Islamic Azad University, Isfahan, Iran.

Research and Professional Experience:

Genetics, System biology, Bioinformatics, Drug Design, Drug Delivery, Non-Coding RNAs, Cancer.

Professional Appointments: Researcher.

Honors:

Best Poster in Royan International twin congress, Reproductive Biomedicine & Stem Cell. Protective approaches of Fraxinus excelsior compounds on the Implantation based infertility via bioinformatics and chemoinformatic analysis.

Publications from the Last 3 Years:

[1] Fatemeh Hajibabaie, Shirin Kouhpayeh, Mina Mirian, Ilnaz Rahimmanesh, Maryam Boshtam, Ladan Sadeghian, Azam Gheibi, Hossein Khanahmad and Laleh Shariati. (2019). MicroRNAs as the actors in the atherosclerosis scenario. *Journal of Physiology and Biochemistry*.
[2] Navid Abedpoor, Farzaneh Taghian, Fatemeh Hajibabaie. (2022). Physical activity ameliorates the function of organs via adipose tissue in metabolic diseases. *Acta histochemical*.
[3] Fatemeh Hajibabaie, Navid Abedpoor, Nazanin Asareh, Mohammad Amin Tabatabaiefar, Ali Zarrabi and Laleh Shariati. (2022). A cocktail of microRNAs as

an advance diagnostic signature in stomach-colorectal cancers hallmarks incidence: a systematic review. *Personal Medicine.*

[4] Navid Abedpoor, Farzaneh Taghian, Fatemeh Hajibabaie. (2022). Cross Brain-Gut Analysis Highlighted Hub Genes and LncRNAs Networks Differentially Modified During Leucine Consumption and Endurance Exercise in Mice with Depression Like Behaviors. *Molecular Neurobiology.*

[5] Fatemeh Hajibabaie, Farzaneh Taghian, Navid Abedpoor, Kamran Safavi. (2023). A Cocktail of Polyherbal Bioactive Compounds and Regular Mobility Training as Senolytic Approaches in Age-dependent Alzheimer's: the In Silico Analysis, Lifestyle Intervention in Old Age. *Journal of Molecular Neuroscience.*

[6] Fatemeh Hajibabaie, Navid Abedpoor, Kamran Safavi, Farzaneh Taghian. (2022). Natural remedies medicine derived from flaxseed (Secoisolariciresinol diglucoside, lignans, and α-linolenic acid) improve network targeting efficiency of diabetic heart conditions based on computational chemistry techniques and pharmacophore modeling. *Journal of Food Biochemistry.*

[7] Fatemeh Hajibabaie, Faranak Aali, Navid Abedpoor. (2022). Pathomechanisms of non-coding RNAs and hub genes related to the oxidative stress in diabetic complications. *F1000Research.*

[8] Fatemeh Hajibabaie, Navid Abedpoor. (2023). The Importance of Hub Genes and Genetically and Epigenetically Modulators as Potent Biomarkers in the Prognosis, Diagnosis, and Therapeutic Monitoring of Colorectal Cancer. *Horizons in Cancer Research.* Volume 85. Nova Science Publishers

Chapter 4

Linolenic Acid Could Modulate the Pathomechanism Related to Colorectal Cancer

Navid Abedpoor[1],*
Fatemeh Hajibabaie[1,2]
Mohammad-Sajad Zare[3,4]
and Kamran Safavi[5]

[1]Department of Physiology, Medicinal Plants Research Center,
Isfahan (Khorasgan) Branch, Islamic Azad University, Isfahan, Iran
[2]Department of Biology, Faculty of Basic Sciences, Shahrekord Branch,
Islamic Azad University, Shahrekord, Iran
[3]Department of Plant and Animal Biology,
Faculty of Biological Sciences and Technology, University of Isfahan, Isfahan, Iran
[4]Iranian Cancer Control Center (MACSA), Isfahan, Iran
[5]Department of Physiology, Medicinal Plants Research Center,
Isfahan (Khorasgan) Branch, Islamic Azad University, Isfahan, Iran

Abstract

Unhealthy lifestyles and modern society cause increasing non-communicable diseases such as cancer, diabetes, and cardiovascular diseases. Growing evidence has demonstrated that dietary factors such as a high-fat diet, advanced glycation end products, and high sugar significantly enhanced gastrointestinal disorders, including inflammatory bowel disease, irritable bowel syndrome, constipation, and colorectal cancer. Colorectal cancer (CRC) is recognized as the most common cancer diagnosed in men and women. In addition, environmental factors such as smoking, a high-fat diet, a Western diet,

---

* Corresponding Author's Email: Abedpoor.navid@yahoo.com, Abedpoor.navid@gmail.com.

In: Properties and Uses of Linolenic Acid
Editor: Calvin S. Willmon
ISBN: 979-8-89113-201-6
© 2023 Nova Science Publishers, Inc.

and a sedentary lifestyle could trigger colorectal cancer. Hence, halting and managing CRC has been essential for public health priority. Data mining and systematic evidence have indicated that nutritional factors have been revealed to play a vital role in the inhibition of CRC. Nutritional enriched n-3 polyunsaturated fatty acids (PUFAs), including alpha-linolenic acid (ALA), have antineoplastic effects in several pathomechanisms. Interestingly, epidemiological evidence has described the protective effect of ALA on obesity-related diseases such as inflammatory bowel disease, colitis, and colorectal cancer. Studies have shown that ALA might activate AMPK/SIRT1, modulate cyclooxygenase activity, and suppress the nuclear factor-κB (NF-κB). Moreover, nutritional-enriched ALA upregulates the novel anti-inflammatory lipid mediators such as resolvins, protectins, and maresins. On the other hand, there are controversial results in which there is no relationship between ALA intake and CRC. Therefore, this chapter will explain how linolenic acid could modulate the pathomechanism related to colorectal cancer.

Introduction

The third most frequent disease in the world, colorectal cancer (CRC), is also the second most common cancer in both women and men. CRC is second in mortality but third in incidence (Sung et al. 2021). Numerous nations in South America, south-central Asia, and southeastern and eastern Europe have seen increased incidence rates. In addition, CRC is recognized as the most significant cause of cancer-related deaths worldwide (Sung et al. 2021). In 2020, around 150,000 cases of CRC and over 50,000 fatalities were reported in the United States (Sung et al. 2021; Siegel et al. 2020).

Growing studies indicated that processed or red meat consumption, obesity, lack of physical activity, westernization of eating habits, and drinking are risk factors, but proper intake of whole fiber, dairy products, and grains lower the risk (Karpisheh et al. 2019). Diet is closely associated with colon carcinogenesis (Ullman and Itzkowitz 2011; Han et al. 2022). It is known that inflammation generated by fatty acid metabolism due to lipid overload, alcohol misuse, and excessive consumption of processed meat raises the risk of tumor growth (Karpisheh et al. 2019; Kim and Park 2003).

Primary prevention is an integral approach to lowering the growing global burden of CRC (Clinton, Giovannucci, and Hursting 2020; Han et al. 2022). Most occurrences of colorectal cancer have been linked to environmental alterations rather than inherited genes (Clinton, Giovannucci, and Hursting

2020). IBD (Inflammatory bowel disease) is a significant colon cancer risk factor. Colitis-related cancer is a form of CRC associated with IBD; it is challenging to treat and has a high death rate (Terzić et al. 2010). In addition, CRC displays constitutive activation of vital transcription factors, such as STAT3 and NF-κB, that impact the tumor microenvironment and tumor cells (Yu, Pardoll, and Jove 2009). Moreover, studies demonstrated that these signaling pathways could integrate roles in cancer-promoting inflammation (Yu, Pardoll, and Jove 2009).

Moreover, inflammation is related to tumor survival, cell proliferation, chemoresistance, metastasis, and angiogenesis. Inflammation may also impair the efficacy of CRC treatments, such as STAT3 inhibitors (Yu, Pardoll, and Jove 2009). NF-κB is also implicated in anti-tumor immune responses; however, STAT3 inhibits NF- κB-mediated anti-tumor immune responses (Zhao and Liu 2018). By redirecting inflammation, STAT3 has been identified as a potential cancer therapy target (Zheng et al. 2017). Tumorigenic JAK/STAT3 signaling activation has also been implicated in advancing colorectal cancer (Liang et al. 2019).

As a result of the low efficacy and significant side effects of standard chemotherapies (Schirrmacher 2019), cancer patients have asked for more effective treatment choices, such as oxaliplatin, to treat colorectal cancer (Mauri et al. 2020). Thus, therapeutic plants and dietary phytochemicals have drawn significant attention due to the widespread perception that these substances are relatively non-toxic because they are edible (Azizian-Farsani et al. 2022; Hajibabaie et al. 2022).

Moreover, n-6 polyunsaturated fatty acids, such as linoleic acid (LA), are implicated in CRC (Liput et al. 2021). Several studies have indicated that cyclooxygenase generates free radicals during prostaglandin E2 production (Upmacis et al. 2022; Lu and Wahl 2005). In contrast, when 15-lipoxygenase-1 metabolizes LA, it activates PPARγ (peroxisome proliferator-activated receptor-γ), inhibits invasion and cancer cell proliferation, and induces apoptosis suppressing tumor growth (Sasaki et al. 2006). Moreover, long-term therapy with LA has been demonstrated to induce quiescence and dormancy in cancer cells, although it can also result in a treatment-resistant phenotype (Ogata et al. 2022). Additionally, arachidonic acid and epoxyeicosatrienoic acid have been linked to the indication of cancer dormancy (Vainio et al. 2011; Panigrahy et al. 2012).

Long incubation periods have been linked to cancer dormancy, which has been linked to late recurrence and metastasis in patients (Wu et al. 2022). For example, in colorectal cancer, cases of recurrence have been recorded several years following tumor removal at the initial site (Wu et al. 2022; Nagata et al. 2020). Dormancy occurs when a tumor stops growing at its initial site or a distant metastatic lesion (Endo and Inoue 2019). Cell dormancy (where cancer cells are in a quiescent state) and tumor mass dormancy (where cell proliferation and cell death are balanced in the tumor) are both referred to as "cancer dormancy" (Gomatou et al. 2021; Tamamouna et al. 2022). The extracellular matrix, the metastatic niche, the hypoxic milieu, and the endoplasmic reticulum stress induce cellular dormancy (Chen and Cubillos-Ruiz 2021).

Tumor dormancy is one reason for resistance to many cancer therapies, and the cells exhibit a stem cell-like phenotype that includes anti-tumor immune suppression, high metastatic potential, and chemoresistance (Emami Nejad et al. 2021; Verma et al. 2023). Simultaneously, cell cycle progression is inhibited in these cells, a distinguishing trait of dormant cancer stem cells (Hen and Barkan 2020). However, the method of cancer cells entering dormancy is not entirely understood. Elucidating the processes of cancer dormancy may aid in improving treatment responsiveness by inducing cancer cells to break dormancy following therapy or immune system escape (Orso et al. 2020). On the other hand, the discovery of techniques to induce ongoing dormancy may contribute to cancer control to avoid clinical recurrence (Orso et al. 2020). It is known that miRNA-494 is upregulated in malignancies and is implicated in angiogenesis through exosome-mediated transport to vascular endothelial cells (Orso et al. 2020; Ogata et al. 2022).

In contrast, in cancer cells, miRNA-494 has been described to target PGC-1a and MYC (Ma et al. 2022; Shams et al. 2020). PGC-1a and MYC are critical regulatory proteins for oxidative phosphorylation and glycolysis, respectively, indicating that miRNA-494 plays a crucial role in cancer cell energy metabolism (Ogata et al. 2022). In addition, Ruiko Ogata and colleagues have indicated that linoleic acid may suppress the quiescence in colorectal cancer cell line CT26 (Ogata et al. 2022). Based on this study, linoleic acid significantly decreased energy production, glycolysis, and oxidative phosphorylation. Moreover, Ogata has demonstrated that the expression level of the MycC and Pgc1a declined in colorectal cancer cell line CT26 by linoleic acid. Interestingly, the expression level of the miR-494 was enhanced in colorectal cancer cell line CT26 by linoleic acid (Ogata et al. 2022).

Factors of diet have been proven to play a significant effect in preventing colorectal cancer (Montalban-Arques and Scharl 2019). In animal models and cellular, polyunsaturated fatty acids (PUFAs, n-3) were shown to be involved in the multiple biological mechanisms underlying the anti-cancer effects (Calviello, Serini, and Piccioni 2007; Yu, Pardoll, and Jove 2009). Alpha-linolenic acid can suppress nuclear factor-B (NF-κB), activate AMPK/SIRT1, modulate cyclooxygenase activity, and upregulate the novel anti-inflammatory lipid (Abdou et al. 2021; Terzić et al. 2010). As a plant-based n-3 PUFA, alpha-linolenic acid can be produced from vegetable oils (Gibson, Muhlhausler, and Makrides 2011). According to population-based epidemiological studies, ALA protects against obesity-related illnesses such as diabetes and cardiovascular disease (Hajibabaie et al. 2022). However, the relationships between ALA in diet and CRC risk were inconsistent in a number of past cohorts, and two previous meta-analyses indicated a null connection assessment (Guldiken et al. 2018; Heo 2009). Given the possibility of measurement error or report bias in most observational cohorts employing dietary questionnaires to estimate ALA intake, it was challenging to determine the intake of particular fatty acids (Huang et al. 2019). In addition, there may be a perturbation of gut microbiota in CRC-vulnerable patients, which may have led to an overestimation of the actual level of ALA in vivo (Liu et al. 2013; Lee et al. 2008). In one study, the levels of ALA in adipose tissue (AT) were inversely linked with the incidence of colorectal cancer (Park and Park 2022). However, in other studies circulating ALA was inversely associated with colon cancer but not rectum cancer (Li et al. 2014) and had no association with CRC risk (Huynh et al. 2019). We indicated the protein-protein interactions involved in the multiple biological mechanisms underlying the anti-cancer effects (Figure 1, Table 1).

Based on the data mining, numerous pathomechanisms identify as ALA's preventive effect against the development of CRC.

1. ALA is a crucial precursor of long-chain n-3 fatty acids in vivo, which can be converted to eicosapentaenoic acid (EPA), docosahexaenoic acid (DHA), and docosapentaenoic acid (DPA) (Burns-Whitmore et al. 2019; Yang et al. 2014). The limited transformation of ALA to marine n-3 PUFAs may have an inhibitory effect on the development of CRC (Burns-Whitmore et al. 2019).
2. As a plant-derived n-3 fatty acid, ALA could influence the activity of cyclooxygenases (COX) and prevent tumor growth by reducing n-6 PUFA-derived 2-series prostaglandin (PGE2) and enhancing n-3 family generated 3-series prostaglandin (PGE3) (Poole et al. 2007).

Table 1. The genes with high degree, betweenness centrality, closeness centrality, and Clustering Coefficient involved in the anti-cancer effects

Symbole genes	Degree	Betweenness Centrality	Closeness Centrality	Clustering Coefficient
ATM	28	0.01563647	0.6627907	0.57936508
SIRT1	35	0.03486118	0.72151899	0.51596639
NFKBIA	26	0.00827974	0.64772727	0.64615385
AURKA	21	0.01486737	0.60638298	0.61904762
KAT2A	21	0.00526487	0.58762887	0.66666667
SKP1	22	0.0069179	0.6	0.61904762
RPA1	10	3.35E-04	0.53271028	0.88888889
MDM2	31	0.0197009	0.68674699	0.55698925
BARD1	11	0.001232	0.52777778	0.74545455
CDH1	19	0.00345085	0.6	0.76608187
CREBBP	36	0.02568453	0.73076923	0.53492063
EP300	43	0.06136833	0.8028169	0.44629014
KAT2B	28	0.01299599	0.64772727	0.59259259
CBL	14	0.00932776	0.55339806	0.56043956
HDAC4	23	0.00454094	0.61290323	0.71541502
STAT3	34	0.05201856	0.7125	0.51158645
EGF	22	0.03206622	0.61956522	0.53679654
CDK2	32	0.02169992	0.69512195	0.53024194
TP53	50	0.14097287	0.890625	0.37469388
CCNA2	23	0.00732822	0.62637363	0.67193676
SKP2	17	0.00253054	0.58762887	0.78676471
EGFR	32	0.05857379	0.69512195	0.45967742
HDAC3	22	0.00634167	0.60638298	0.66666667
CEBPB	26	0.01035973	0.64772727	0.64307692
NPM1	17	0.00355303	0.58762887	0.71323529
BCL2L1	19	0.00365624	0.6	0.69590643
FBXW7	23	0.01141477	0.62637363	0.60079051
MYC	45	0.06190863	0.82608696	0.44040404
WRN	10	5.44E-04	0.53271028	0.82222222
RBX1	22	0.00931397	0.61290323	0.57575758
NFKB1	22	0.00504292	0.61956522	0.67965368
CKS1B	9	0	0.52293578	1
RASA1	4	5.07E-04	0.4488189	0.66666667
UBE3A	11	0.00329942	0.54285714	0.58181818
RELB	14	0.00122415	0.55882353	0.74725275
PPARG	27	0.01646504	0.65517241	0.58974359
REL	14	7.43E-04	0.57	0.84615385
STAT2	9	0.00396931	0.54285714	0.75
HBEGF	9	0.0022717	0.53271028	0.77777778
PLCG1	8	0.00153477	0.47107438	0.75
EREG	6	0	0.44186047	1
TGFA	9	0.00378935	0.54285714	0.72222222
SMAD2	23	0.00588067	0.62637363	0.6916996

Symbole genes	Degree	Betweenness Centrality	Closeness Centrality	Clustering Coefficient
TBL1X	15	0.00220048	0.57	0.74285714
MYOD1	16	9.33E-04	0.57	0.875
PPARGC1A	28	0.59782	0.633333	0.019797

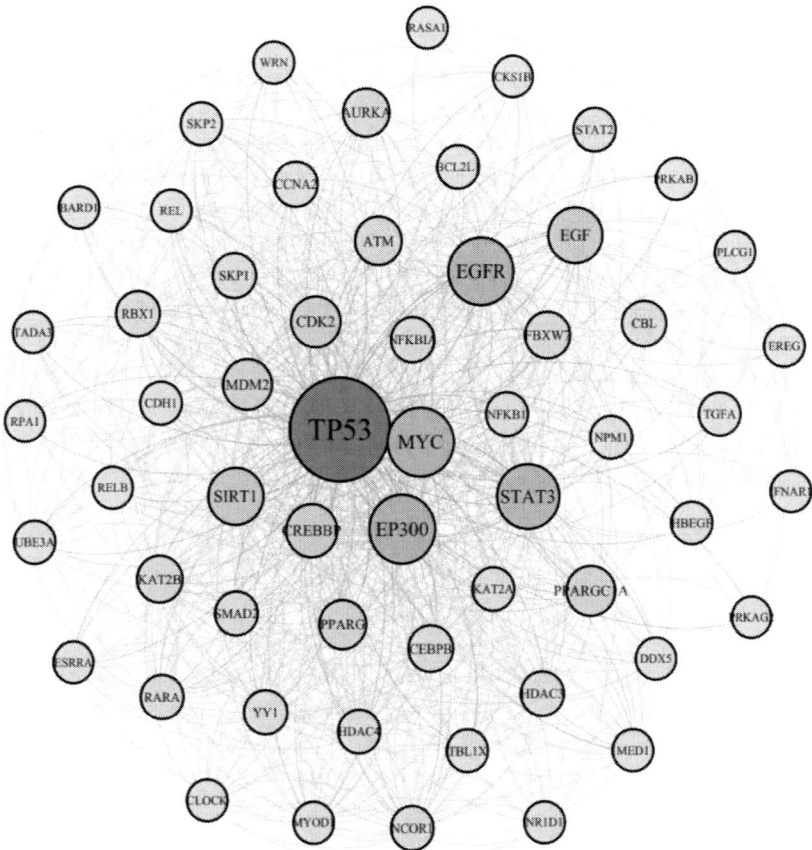

Figure 1. The protein- protein intraction involved in the multiple biological mechanisms underlying the anti-cancer effects.

3. ALA inhibited the inflammatory phenotype of M1-like macrophages, consequently decreasing the expression levels of pro-inflammatory markers like IL-1β, TNF-α, MCP-1, and IL-6 in human THP-1 cells (Pauls et al. 2018).
4. ALA may have independently regulated the apoptotic mechanism and NF-κB signaling pathway associated with the inflammatory response

to control tumor proliferation, migrations, and invasions (Pauls et al. 2018; González-Fernández, Ortea, and Guil-Guerrero 2020).

Table 2. The effect of the ALA on the colorectal cancer

Study	Animal/cell line	Approaches
(Sasaki et al. 2006)	BALB/c mice	Reduced metastatic foci, stimulation PPRn.
(Nichenametla, South, and Exon 2004)	Sprague-Dawley Rat	Prevent aberrant crypts, and improve killer cell activity.
(Kim and Park 2003)	Sprague-Dawley Rat	Decreased tumor size and enhanced apoptosis.
(Petrik et al. 2000)	Sprague-Dawley Rat	Declined tumor size and enhanced apoptosis.
(Petrik et al. 2000)	Apc(min/+) mice	No effect.
(Soel et al. 2007)	BALB/c mice	Suppress cancer cell migration and decreased pulmonary nodules
(Shiraishi et al. 2010)	Sprague-Dawley Rat	Enhanced apoptosis, and induced Caspase-3
(Mandir and Goodlad 2008)	Apc(min/+) mice	Amplified apoptotic, decreased tumor size
(Han J Cho et al. 2003)	HT-29	Suppress proliferation via ErbB3 signaling pathway.
(Palombo et al. 2002)	HT-29, PC-3, MIP-101	Halt the caspase signaling pathway.
(Han Jin Cho et al. 2005)	HT-29	Suppress proliferation via ErbB3 signaling pathway.
(Palombo et al. 2002)	Caco-2	Suppress proliferation Reduced insulin-like growth factor II (IGF-II)
(Lim et al. 2005)	HT-29	Enhanced cell cycle arrest at G0/G1 phase
(Lampen et al. 2005)	Caco-2	Suppress proliferation

The data indicated that the PGE2 as an inflammatory mediator significantly decreased by consuming the ALA. Moreover, ALA can regulate the protein Bax/Bcl-2 as an apoptosis signaling pathway (Mandir and Goodlad 2008). Another study showed that consuming the ALA may decline tumor size and enhance apoptosis. In addition, the PPARγ (Peroxisome proliferator-activated receptor-γ) expression level is regulated by the consumption of the ALA in BALB/c mice induced by azoxymethane (Sasaki et al. 2006). Based on these data polyp PPARγ signaling pathway is related to the diameter of the intestinal neoplasia (Sasaki et al. 2006). Growing evidence indicated that ALA could inhibit the colorectal cancer tumors involved improved killer cell activity (Nichenametla, South, and Exon 2004), enhanced apoptosis (Petrik et al. 2000; Kim and Park 2003), decreased pulmonary nodules (Soel et al. 2007), induced Caspase-3 (Shiraishi et al. 2010; Palombo et al. 2002), suppress proliferation via ErbB3 signaling pathway (Han J Cho et al. 2003; Han Jin Cho et al. 2005), reduced insulin-like growth factor II (IGF-II) (Palombo et al.

2002), suppress proliferation (Lampen et al. 2005), and enhanced cell cycle arrest at G0/G1 phase (Lim et al. 2005). Table 2 summarizes the study, which indicated the effect of the consumption of ALA on colorectal cancer.

Conclusion

Nutritional-enriched ALA upregulates the novel anti-inflammatory lipid mediators such as resolvins, protectins, and maresins. Moreover, ALA decreased the expression levels of pro-inflammatory markers like IL-1β, TNF-α, MCP-1, and IL-6. Hence, ALA might inhibit the inflammatory phenotype of M1-like macrophages.

References

Abdou, Rabab M, Walaa H El-Maadawy, Marwa Hassan, Riham S El-Dine, Tarek Aboushousha, Nebal D El-Tanbouly, and Aly M El-Sayed. 2021. "Nephroprotective activity of Aframomum melegueta seeds extract against diclofenac-induced acute kidney injury: A mechanistic study." *Journal of Ethnopharmacology* 273: 113939.

Azizian-Farsani, Fatemeh, Navid Abedpoor, Maryam Derakhshan, Mohammad Hossein Nasr-Esfahani, Mohammad Hasan Sheikhha, and Kamran Ghaedi. 2022. "Protective Effects of the Combination of the Herbal Compound Against Inflammation Related to Obesity and Colitis Induced by Diet in Mice." *Iranian journal of diabetes and obesity*.

Burns-Whitmore, Bonny, Erik Froyen, Celine Heskey, Temetra Parker, and Gregorio San Pablo. 2019. "Alpha-linolenic and linoleic fatty acids in the vegan diet: do they require dietary reference intake/adequate intake special consideration?" *Nutrients* 11 (10): 2365.

Calviello, Gabriella, Simona Serini, and Elisabetta Piccioni. 2007. "n-3 polyunsaturated fatty acids and the prevention of colorectal cancer: molecular mechanisms involved." *Current medicinal chemistry* 14 (29): 3059-3069.

Chen, Xi, and Juan R Cubillos-Ruiz. 2021. "Endoplasmic reticulum stress signals in the tumour and its microenvironment." *Nature Reviews Cancer* 21 (2): 71-88.

Cho, Han J, Woo K Kim, Eun J Kim, Kyeong C Jung, Soochul Park, Hyun S Lee, Angela L Tyner, and Jung HY Park. 2003. "Conjugated linoleic acid inhibits cell proliferation and ErbB3 signaling in HT-29 human colon cell line." *American Journal of Physiology-Gastrointestinal and Liver Physiology* 284 (6): G996-G1005.

Cho, Han Jin, Woo Kyoung Kim, Jae In Jung, Eun Ji Kim, Soon Sung Lim, Dae Young Kwon, and Jung Han Yoon Park. 2005. "Trans-10, cis-12, not cis-9, trans-11, conjugated linoleic acid decreases ErbB3 expression in HT-29 human colon cancer cells." *World Journal of Gastroenterology: WJG* 11 (33): 5142.

Clinton, Steven K, Edward L Giovannucci, and Stephen D Hursting. 2020. "The world cancer research fund/American institute for cancer research third expert report on diet, nutrition, physical activity, and cancer: impact and future directions." *The Journal of nutrition* 150 (4): 663-671.

Emami Nejad, Asieh, Simin Najafgholian, Alireza Rostami, Alireza Sistani, Samaneh Shojaeifar, Mojgan Esparvarinha, Reza Nedaeinia, Shaghayegh Haghjooy Javanmard, Marjan Taherian, and Mojtaba Ahmadlou. 2021. "The role of hypoxia in the tumor microenvironment and development of cancer stem cell: a novel approach to developing treatment." *Cancer Cell International* 21 (1): 1-26.

Endo, Hiroko, and Masahiro Inoue. 2019. "Dormancy in cancer." *Cancer science* 110 (2): 474-480.

Gibson, Robert A, Bev Muhlhausler, and Maria Makrides. 2011. "Conversion of linoleic acid and alpha-linolenic acid to long-chain polyunsaturated fatty acids (LCPUFAs), with a focus on pregnancy, lactation and the first 2 years of life." *Maternal & child nutrition* 7: 17-26.

Gomatou, Georgia, Nikolaos Syrigos, Ioannis A Vathiotis, and Elias A Kotteas. 2021. "Tumor dormancy: implications for invasion and metastasis." *International journal of molecular sciences* 22 (9): 4862.

González-Fernández, María José, Ignacio Ortea, and José Luis Guil-Guerrero. 2020. "α-Linolenic and γ-linolenic acids exercise differential anti-tumor effects on HT-29 human colorectal cancer cells." *Toxicology Research* 9 (4): 474-483.

Guldiken, Burcu, Gulay Ozkan, Gizem Catalkaya, Fatma Duygu Ceylan, Ipek Ekin Yalcinkaya, and Esra Capanoglu. 2018. "Phytochemicals of herbs and spices: Health versus toxicological effects." *Food and Chemical Toxicology* 119: 37-49.

Hajibabaie, Fatemeh, Navid Abedpoor, Kamran Safavi, and Farzaneh Taghian. 2022. "Natural remedies medicine derived from flaxseed (secoisolariciresinol diglucoside, lignans, and α-linolenic acid) improve network targeting efficiency of diabetic heart conditions based on computational chemistry techniques and pharmacophore modeling." *Journal of Food Biochemistry*: e14480.

Han, Sanghee, Hail Kim, Min Young Lee, Junhee Lee, Kwang Seok Ahn, In Jin Ha, and Seok-Geun Lee. 2022. "Anti-Cancer Effects of a New Herbal Medicine PSY by Inhibiting the STAT3 Signaling Pathway in Colorectal Cancer Cells and Its Phytochemical Analysis." *International Journal of Molecular Sciences* 23 (23): 14826.

Hen, Omri, and Dalit Barkan. 2020. "Dormant disseminated tumor cells and cancer stem/progenitor-like cells: Similarities and opportunities." *Seminars in Cancer Biology*.

Heo, Jun. 2009. "Dongui-Bogam: Treasured Mirror of Eastern Medicine." *Chin Young, Ministry of Health & Welfare, Jongno-gu, Seoul, Korea*.

Huang, Si-zhou, Wang-yu Liu, Yue Huang, A-ling Shen, Li-ya Liu, and Jun Peng. 2019. "Patrinia scabiosaefolia inhibits growth of 5-FU-resistant colorectal carcinoma cells via induction of apoptosis and suppression of AKT pathway." *Chinese journal of integrative medicine* 25: 116-121.

Huynh, Jennifer, Ashwini Chand, Daniel Gough, and Matthias Ernst. 2019. "Therapeutically exploiting STAT3 activity in cancer—using tissue repair as a road map." *Nature Reviews Cancer* 19 (2): 82-96.

Karpisheh, Vahid, Afshin Nikkhoo, Mohammad Hojjat-Farsangi, Afshin Namdar, Gholamreza Azizi, Ghasem Ghalamfarsa, Gholamabas Sabz, Mehdi Yousefi, Bahman Yousefi, and Farhad Jadidi-Niaragh. 2019. "Prostaglandin E2 as a potent therapeutic target for treatment of colon cancer." *Prostaglandins & other lipid mediators* 144: 106338.

Kim, Kyung-Hee, and Hyun-Suh Park. 2003. "Dietary supplementation of conjugated linoleic acid reduces colon tumor incidence in DMH-treated rats by increasing apoptosis with modulation of biomarkers." *Nutrition* 19 (9): 772-777.

Lampen, A, M Leifheit, J Voss, and H Nau. 2005. "Molecular and cellular effects of cis-9, trans-11-conjugated linoleic acid in enterocytes: effects on proliferation, differentiation, and gene expression." *Biochimica et Biophysica Acta (BBA)-Molecular and Cell Biology of Lipids* 1735 (1): 30-40.

Lee, Ming-Yi, Huan-You Lin, Faiwen Cheng, Wenchang Chiang, and Yueh-Hsiung Kuo. 2008. "Isolation and characterization of new lactam compounds that inhibit lung and colon cancer cells from adlay (Coix lachryma-jobi L. var. ma-yuen Stapf) bran." *Food and Chemical Toxicology* 46 (6): 1933-1939.

Li, Bohui, Weiyang Tao, Chunli Zheng, Piar Ali Shar, Chao Huang, Yingxue Fu, and Yonghua Wang. 2014. "Systems pharmacology-based approach for dissecting the addition and subtraction theory of traditional Chinese medicine: an example using Xiao-Chaihu-Decoction and Da-Chaihu-Decoction." *Computers in Biology and Medicine* 53: 19-29.

Liang, Chaojie, Tuanjie Zhao, Haijun Li, Fucheng He, Xin Zhao, Yuan Zhang, Xi Chu, Chunlan Hua, Yunhui Qu, and Yu Duan. 2019. "Long non-coding RNA ITIH4-AS1 accelerates the proliferation and metastasis of colorectal cancer by activating JAK/STAT3 signaling." *Molecular Therapy-Nucleic Acids* 18: 183-193.

Lim, Do Y, Angela L Tyner, Jae-Bong Park, Jae-Yong Lee, Yung H Choi, and Jung HY Park. 2005. "Inhibition of colon cancer cell proliferation by the dietary compound conjugated linoleic acid is mediated by the CDK inhibitor p21CIP1/WAF1." *Journal of cellular physiology* 205 (1): 107-113.

Liput, Kamila P, Adam Lepczyński, Magdalena Ogłuszka, Agata Nawrocka, Ewa Poławska, Agata Grzesiak, Brygida Ślaska, Chandra S Pareek, Urszula Czarnik, and Mariusz Pierzchała. 2021. "Effects of dietary n–3 and n–6 polyunsaturated fatty acids in inflammation and cancerogenesis." *International journal of molecular sciences* 22 (13): 6965.

Liu, Liya, Aling Shen, Youqin Chen, Lihui Wei, Jiumao Lin, Thomas J Sferra, Zhenfeng Hong, and Jun Peng. 2013. "Patrinia scabiosaefolia induces mitochondrial-dependent apoptosis in a mouse model of colorectal cancer." *Oncology Reports* 30 (2): 897-903.

Lu, Yunbiao, and Larry M Wahl. 2005. "Oxidative stress augments the production of matrix metalloproteinase-1, cyclooxygenase-2, and prostaglandin E2 through enhancement of NF-κB activity in lipopolysaccharide-activated human primary monocytes." *The Journal of Immunology* 175 (8): 5423-5429.

Ma, Lunkun, Ankit Gilani, Qian Yi, and Liling Tang. 2022. "MicroRNAs as Mediators of Adipose Thermogenesis and Potential Therapeutic Targets for Obesity." *Biology* 11 (11): 1657.

Mandir, N, and RA Goodlad. 2008. "Conjugated linoleic acids differentially alter polyp number and diameter in the Apcmin/+ mouse model of intestinal cancer." *Cell proliferation* 41 (2): 279-291.

Mauri, G, S Arena, S Siena, A Bardelli, and A Sartore-Bianchi. 2020. "The DNA damage response pathway as a land of therapeutic opportunities for colorectal cancer." *Annals of Oncology* 31 (9): 1135-1147.

Montalban-Arques, Ana, and Michael Scharl. 2019. "Intestinal microbiota and colorectal carcinoma: Implications for pathogenesis, diagnosis, and therapy." *EBioMedicine* 48: 648-655.

Nagata, Hiroshi, Kazushige Kawai, Keisuke Hata, Toshiaki Tanaka, Hiroaki Nozawa, and Soichiro Ishihara. 2020. "Laparoscopic surgery for T4 colon cancer: a risk factor for peritoneal recurrences?" *Surgery* 168 (1): 119-124.

Nichenametla, Sailendra N, Elizabeth H South, and Jerry H Exon. 2004. "Interaction of conjugated linoleic acid, sphingomyelin, and butyrate on formation of colonic aberrant crypt foci and immune functions in rats." *Journal of Toxicology and Environmental Health, Part A* 67 (6): 469-481.

Ogata, Ruiko, Shiori Mori, Shingo Kishi, Rika Sasaki, Naoya Iwata, Hitoshi Ohmori, Takamitsu Sasaki, Yukiko Nishiguchi, Chie Nakashima, and Kei Goto. 2022. "Linoleic acid upregulates microrna-494 to induce quiescence in colorectal cancer." *International Journal of Molecular Sciences* 23 (1): 225.

Orso, Francesca, Lorena Quirico, Daniela Dettori, Roberto Coppo, Federico Virga, Livia C Ferreira, Camilla Paoletti, Désirée Baruffaldi, Elisa Penna, and Daniela Taverna. 2020. "Role of miRNAs in tumor and endothelial cell interactions during tumor progression." Seminars in Cancer Biology.

Palombo, John D, Aniruddha Ganguly, Bruce R Bistrian, and Michael P Menard. 2002. "The antiproliferative effects of biologically active isomers of conjugated linoleic acid on human colorectal and prostatic cancer cells." *Cancer letters* 177 (2): 163-172.

Panigrahy, Dipak, Matthew L Edin, Craig R Lee, Sui Huang, Diane R Bielenberg, Catherine E Butterfield, Carmen M Barnés, Akiko Mammoto, Tadanori Mammoto, and Ayala Luria. 2012. "Epoxyeicosanoids stimulate multiorgan metastasis and tumor dormancy escape in mice." *The Journal of clinical investigation* 122 (1): 178-191.

Park, Hyun-Ji, and Shin-Hyung Park. 2022. "Root Bark of Morus Alba L. Induced p53-Independent Apoptosis in Human Colorectal Cancer Cells by Suppression of STAT3 Activity." *Nutrition and Cancer* 74 (5): 1837-1848.

Pauls, Samantha D, Lisa A Rodway, Tanja Winter, Carla G Taylor, Peter Zahradka, and Harold M Aukema. 2018. "Anti-inflammatory effects of α-linolenic acid in M1-like macrophages are associated with enhanced production of oxylipins from α-linolenic and linoleic acid." *The Journal of nutritional biochemistry* 57: 121-129.

Petrik, Melissa B Hansen, Michael F McEntee, Benjamin T Johnson, Mark G Obukowicz, and Jay Whelan. 2000. "Highly unsaturated (n-3) fatty acids, but not α-linolenic, conjugated linoleic or γ-linolenic acids, reduce tumorigenesis in Apc Min/+ mice." *The Journal of nutrition* 130 (10): 2434-2443.

Poole, Elizabeth M, Jeannette Bigler, John Whitton, Justin G Sibert, Richard J Kulmacz, John D Potter, and Cornelia M Ulrich. 2007. "Genetic variability in prostaglandin synthesis, fish intake and risk of colorectal polyps." *Carcinogenesis* 28 (6): 1259-1263.

Sasaki, Takamitsu, Kiyomu Fujii, Kazuhiro Yoshida, Hideo Shimura, Tomonori Sasahira, Hitoshi Ohmori, and Hiroki Kuniyasu. 2006. "Peritoneal metastasis inhibition by linoleic acid with activation of PPARγ in human gastrointestinal cancer cells." *Virchows Archiv* 448: 422-427.

Schirrmacher, Volker. 2019. "From chemotherapy to biological therapy: A review of novel concepts to reduce the side effects of systemic cancer treatment." *International journal of oncology* 54 (2): 407-419.

Shams, Roshanak, Hamid Asadzadeh Aghdaei, Ali Behmanesh, Amir Sadeghi, Mohammadareza Zali, Sina Salari, and Jose M Padron. 2020. "MicroRNAs targeting MYC expression: trace of hope for pancreatic cancer therapy. a systematic review." *Cancer management and research* 12: 2393.

Shiraishi, Ryosuke, Ryuichi Iwakiri, Takehiro Fujise, Tsukasa Kuroki, Takashi Kakimoto, Tooru Takashima, Yasuhisa Sakata, Seiji Tsunada, Yutaka Nakashima, and Teruyoshi Yanagita. 2010. "Conjugated linoleic acid suppresses colon carcinogenesis in azoxymethane-pretreated rats with long-term feeding of diet containing beef tallow." *Journal of gastroenterology* 45: 625-635.

Siegel, Rebecca L, Kimberly D Miller, Ann Goding Sauer, Stacey A Fedewa, Lynn F Butterly, Joseph C Anderson, Andrea Cercek, Robert A Smith, and Ahmedin Jemal. 2020. "Colorectal cancer statistics, 2020." *CA: a cancer journal for clinicians* 70 (3): 145-164.

Soel, So Mi, Ok Sook Choi, Myung Hee Bang, Jung Han Yoon Park, and Woo Kyoung Kim. 2007. "Influence of conjugated linoleic acid isomers on the metastasis of colon cancer cells in vitro and in vivo." *The Journal of nutritional biochemistry* 18 (10): 650-657.

Sung, Hyuna, Jacques Ferlay, Rebecca L Siegel, Mathieu Laversanne, Isabelle Soerjomataram, Ahmedin Jemal, and Freddie Bray. 2021. "Global cancer statistics 2020: GLOBOCAN estimates of incidence and mortality worldwide for 36 cancers in 185 countries." *CA: a cancer journal for clinicians* 71 (3): 209-249.

Tamamouna, Vasilia, Evangelia Pavlou, Christiana M Neophytou, Panagiotis Papageorgis, and Paul Costeas. 2022. "Regulation of Metastatic Tumor Dormancy and Emerging Opportunities for Therapeutic Intervention." *International Journal of Molecular Sciences* 23 (22): 13931.

Terzić, Janoš, Sergei Grivennikov, Eliad Karin, and Michael Karin. 2010. "Inflammation and colon cancer." *Gastroenterology* 138 (6): 2101-2114. e5.

Ullman, Thomas A, and Steven H Itzkowitz. 2011. "Intestinal inflammation and cancer." *Gastroenterology* 140 (6): 1807-1816. e1.

Upmacis, Rita K, Wendy L Becker, Donna M Rattendi, Raven S Bell, Kelsey D Jordan, Shayan Saniei, and Elena Mejia. 2022. "Analysis of Sex-Specific Prostanoid Production Using a Mouse Model of Selective Cyclooxygenase-2 Inhibition." *Biomarker Insights* 17: 11772719221142151.

Vainio, Paula, Santosh Gupta, Kirsi Ketola, Tuomas Mirtti, John-Patrick Mpindi, Pekka Kohonen, Vidal Fey, Merja Perälä, Frank Smit, and Gerald Verhaegh. 2011. "Arachidonic acid pathway members PLA2G7, HPGD, EPHX2, and CYP4F8 identified as putative novel therapeutic targets in prostate cancer." *The American journal of pathology* 178 (2): 525-536.

Verma, Poornima, Neha Shukla, Shivani Kumari, MS Ansari, Naveen Kumar Gautam, and Girijesh Kumar Patel. 2023. "Cancer stem cell in prostate cancer progression, metastasis and therapy resistance." *Biochimica et Biophysica Acta (BBA)-Reviews on Cancer*: 188887.

Wu, Rongrong, Arya Mariam Roy, Yoshihisa Tokumaru, Shipra Gandhi, Mariko Asaoka, Masanori Oshi, Li Yan, Takashi Ishikawa, and Kazuaki Takabe. 2022. "NR2F1, a tumor dormancy marker, is expressed predominantly in cancer-associated fibroblasts and is associated with suppressed breast cancer cell proliferation." *Cancers* 14 (12): 2962.

Yang, Bo, Feng-Lei Wang, Xiao-Li Ren, and Duo Li. 2014. "Biospecimen long-chain N-3 PUFA and risk of colorectal cancer: a meta-analysis of data from 60,627 individuals." *PLoS One* 9 (11): e110574.

Yu, Hua, Drew Pardoll, and Richard Jove. 2009. "STATs in cancer inflammation and immunity: a leading role for STAT3." *Nature reviews cancer* 9 (11): 798-809.

Zhao, Yingke, and Yue Liu. 2018. "A mechanistic overview of herbal medicine and botanical compounds to target transcriptional factors in Breast cancer." *Pharmacological Research* 130: 292-302.

Zheng, Xiang, Kati Turkowski, Javier Mora, Bernhard Brüne, Werner Seeger, Andreas Weigert, and Rajkumar Savai. 2017. "Redirecting tumor-associated macrophages to become tumoricidal effectors as a novel strategy for cancer therapy." *Oncotarget* 8 (29): 48436.

Biographical Sketch

Name: *Navid Abedpoor*

Affiliation: Department of Sports Physiology, Faculty of Sports Sciences, Isfahan (Khorasgan) Branch, Islamic Azad University, Isfahan, Iran.

Department of Physiology, Medicinal Plants Research Center, Isfahan (Khorasgan) Branch, Islamic Azad University, Isfahan, Iran.

Education: Sport sciences.

Business Address: Isfahan (Khorasgan) Branch, Islamic Azad University, Isfahan, Iran.

Research and Professional Experience: Inflammation, Oxidative Stress, Non coding RNAs, Cancer, Lifestyles.

Professional Appointments: Researcher.

Honors:

- Won the Young Scientist Award in "International Scientist Awards on Engineering, Science, and Medicine." 2020.
- Best Poster in Royan International twin congress, Reproductive Biomedicine & Stem Cell. Protective approaches of Fraxinus excelsior compounds on the Implantation based infertility via bioinformatics and chemoinformatic analysis.

Publications from the Last 3 Years:

1. Fatemeh Azizian-Farsani, Navid Abedpoor, Mohammad Hasan Sheikhha, Ali Osmay Gure, Mohammad Hossein Nasr Esfahani, Kamran Ghaedi. (2020). Receptor for advanced glycation end products acts as a fuel to colorectal cancer development. *Frontiers in Oncology*. IF: 6.5.
2. Fatemeh Azizian-Farsani, Marcin Osuchowski, Navid Abedpoor, Farzad Seyed Forootan, Maryam Derakhshan, Mohammad Hossein Nasr-Esfahani, Mohammad Hasan Sheikhha, Kamran Ghaedi. (2020). Anti-inflammatory and -apoptotic effects of an herbal extract on DSS-induced colitis in mice fed with high AGEs-fat diet. Scientific Reports. *Nutrition & Metabolism*. IF: 4.5.
3. Golbarg Rahimi; Salime Heydari; Bahare Rahimi; Navid Abedpoor; Iman Nicktab; Zahra Safaeinejad; Maryam Peymani; Farzad Seyed Forootan; Zahra Derakhshan; Mohammad Hossein Nasr Esfahani, Kamran Ghaedi. (2020). A combination of herbal compound (SPTC) along with exercise or metformin more efficiently alleviated diabetic complications through down-regulation of stress oxidative pathway upon activating Nrf2-Keap1 axis in AGEs rich diet-induced type 2 diabetic mice. *Nutrition and metabolism*.IF:4.5.
4. Fahimeh Akbarian, Mohsen Rahmani, Marziyeh Tavalaee, Navid Abedpoor, Mozhdeh Taki, Kamran Ghaedi, Mohammad Hossein Nasr-Esfahani. (2021). Effect of different high-fat and AGEs diets in obesity and diabetes-prone C57BL/6 mice on sperm function. *International Journal of Fertility and Sterility*. IF:2.7.
5. Navid Abedpoor, Farzaneh Taghian, Fatemeh Hajibabaie. (2022). Physical activity ameliorates the function of organs via adipose tissue in metabolic diseases. *Acta histochemical*. IF: 2.7.
6. Fatemeh Hajibabaie, Navid Abedpoor, Nazanin Asareh, Mohammad Amin Tabatabaiefar, Ali Zarrabi and Laleh Shariati. (2022). A cocktail of microRNAs as

an advance diagnostic signature in stomach-colorectal cancers hallmarks incidence: a systematic review. *Personal Medicine.*
7. Navid Abedpoor, Farzaneh Taghian, Fatemeh Hajibabaie. (2022). Cross Brain-Gut Analysis Highlighted Hub Genes and LncRNAs Networks Differentially Modified During Leucine Consumption and Endurance Exercise in Mice with Depression Like Behaviors. *Molecular Neurobiology.* IF: 5.5.
8. Navid Abedpoor, Iman Niktab, Mohammad-Hossein Beigi, Masoud Baghi, Fahimeh Arzande, Naeimeh Rezaei, Mohammad-Sajad Zare, Timothy L. Megraw, Hoi-Ying Holman, Amirkianoosh Kiani, Farzad Seyed Forootan, Hossein Baharvand, Mohammad Hossein Nasr Esfahani, Kamran Ghaedi (2022). *Exercise facilitates the browning of fat tissue by up-regulating Irisin receptors.* (Submitted).
9. Maryam Haghparast Azad, Iman Niktab, Shaghayegh Dastjerdi, Navid Abedpoor, Golbarg Rahimi, Zahra Safaeinejad, Maryam Peymani, Farzad Seyed Forootan, Majid Asadi-Shekaari, Mohammad Hossein Nasr Esfahani & Kamran Ghaedi. The combination of endurance exercise and SGTC (Salvia–Ginseng–Trigonella–Cinnamon) ameliorate mitochondrial markers' overexpression with sufficient ATP production in the skeletal muscle of mice fed AGEs-rich high-fat diet. (2022). *Nutrition & Metabolism.*
10. Golnaz Pakravan, Maryam Peymani, Navid Abedpoor, Zahra Safaeinejad, Mehrdad Yadegari, Maryam Derakhshan, Mohammad Hossein Nasr Esfahani, Kamran Ghaedi. Antiapoptotic and anti-inflammatory effects of Pparγ agonist, pioglitazone, reversed Dox-induced cardiotoxicity through mediating of miR-130a downregulation in C57BL/6 mice. (2022). *Journal of Biochemical and Molecular Toxicology.*
11. Fatemeh Azizian-Farsani, Navid Abedpoor, Maryam Derakhshan, Mohammad Hossein Nasr-Esfahani, Mohammad Hasan Sheikhha, Kamran Ghaedi. Protective Effects of the Combination of the Herbal Compound Against Inflammation Related to Obesity and Colitis Induced by Diet in Mice. (2022). *Iranian Journal of Diabetes and Obesity.*
12. Global multi-stakeholder endorsement of the MAFLD definition. (2022). *The Lancet Gastroenterology & Hepatology.* IF: 200.
13. Fatemeh Hajibabaie, Faranak Aali, Navid Abedpoor*. Pathomechanisms of non-coding RNAs and hub genes related to the oxidative stress in diabetic complications. (2022). *F1000Research.*
14. Fatemeh Hajibabaie, Navid Abedpoor*, Kamran Safavi, F. Taghian. Natural remedies medicine derived from flaxseed (secoisolariciresinol diglucoside, lignans, and α-linolenic acid) improve network targeting efficiency of diabetic heart conditions based on computational chemistry techniques and pharmacophore modeling. (2022). *Journal of food biochemistry.*
15. Fatemeh Hajibabaie, Navid Abedpoor*, Farzaneh Taghian, Kamran Safavi. A cocktail of polyherbal bioactive compounds and regular mobility training as senolytic approaches in age-dependent Alzheimer's: The in-silico analysis, lifestyle intervention in old age. (2022). *Journal of Molecular Neuroscience.*

Chapter 5

Natural Remedies that Offset Alpha-Linolenic Acid (ALA) Deficiency in Cardiovascular Hallmarks and Complications

Fatemeh Hajibabaie[1,2]
Navid Abedpoor[1,*]
and Elina Kaviani[3]

[1]Department of Physiology, Medicinal Plants Research Center, Isfahan (Khorasgan) Branch, Islamic Azad University, Isfahan, Iran
[2]Department of Biology, Faculty of Basic Sciences, Shahrekord Branch, Islamic Azad University, Shahrekord, Iran
[3]Isfahan Endocrine and Metabolism Research Center, Isfahan University of Medical Sciences, Isfahan, Iran

Abstract

The growing number of evidence has shown that dietary lipids influence the development of atherosclerosis, cardiac arrest, abnormal heart rhythms, lipid profile, and cardiovascular risk. Coronary heart disease, cerebrovascular disease, rheumatic heart disease, and various disturbances of the blood arteries system and heart functions are all included in the category of cardiovascular diseases (CVDs). An estimated 17.9 million people worldwide lose their lives each year to cardiovascular diseases (CVDs). Inherited and sporadic forms of cardiovascular disease both have their origins in a combination of genetic predisposition and exposure to adverse environmental variables such as sedentary behavior, smoking, and a diet rich in fatty acids. One of the most important aspects of arterial thrombus is atherosclerosis, which

---

[*] Corresponding Author.

In: Properties and Uses of Linolenic Acid
Editor: Calvin S. Willmon
ISBN: 979-8-89113-201-6
© 2023 Nova Science Publishers, Inc.

occurs as a result of plaque buildup and rupture of artery walls in patients with dyslipidemia. By maintaining low-density lipoproteins (LDL) under the endothelium, cholesterol homeostasis was disrupted, activating inflammatory cascades and causing the migration of monocytes to areas of inflammation. All of these things cause cardiac cells to produce more reactive oxygen species (ROS). Some molecular physiologists have hypothesized that chronic inflammation, apoptosis, and oxidative stress play critical roles in contributing to cardiovascular pathogenesis. Due to the obvious evidence showing the beneficial properties of plant-based nutrition and long-chain n-3 fatty acids, there is a great interest to discover more about the effects of the plant-based n-3 fatty acid alpha-linolenic acid (ALA) on cardiovascular disease, and metabolic syndrome. This chapter reviews the existing literature on ALA and gives a brief overview of the many cardiovascular attractive features of ALA.

Introduction

Epidemiology and Risk Factors

Non-communicable diseases (NCDs) are progressively developing into severe hazards to human life and health (Hajibabaie et al., 2020). Moreover, the prevalence of cardiovascular disease (CVD) is continually expanding and becoming a major risk factor impacting human health. It causes 17.5 million deaths annually, accounting for 31 percent of total fatalities worldwide (Yue et al., 2021). As the most prevalent NCDs with significant suffering and mortality annually, heart and blood vessel diseases and associated complications impose a major cost on global public health (Gaidai, Cao, and Loginov 2023); hence CVDs is described as a primary and significant public issue. CVD is a general term in medicine and physiology sciences involving many diseases in the heart, blood vessels, and circulation systems (Laslett et al., 2012). This concept includes high blood pressure (hypertension), ischaemic heart disease (ICD), coronary artery disease (CAD), atherosclerosis, cerebrovascular diseases (stenosis, thrombosis, embolism, hemorrhage, Moya Moya, and stroke), heart failure (HF) and other abnormality in the heart (Gaidai, Cao, and Loginov 2023). Etiological research has mentioned that high blood pressure, abnormal lipid profile, sedentary lifestyle, smoking, diabetes, a dietary regime rich in saturated fat, positive familial history, and inheritance triggered abnormal phenotypic and functional statuses (clinical and preclinical symptoms) in the cardiovascular system. Due to the high prevalence, significant mortality rate, and swift changes in the cardiology field, prognosis, diagnosis, monitoring, therapeutic strategies, and accurate statistics are vital tools for CVDs management (C.W. Tsao et al. 2022a; C. Tsao et al., 2022b).

Inflammation promotes endothelial dysfunction, accumulation of lipids, and a cholesterol-rich atherosclerotic plaque build-up (Hajibabaie et al., 2020). Atherosclerosis, as one of the inflammatory signs of cardiovascular disease, is initiated by vascular injuries and lipid depositions aggregation that monocyte phagocytosis recruitment and establish foam cells (Hoseini et al., 2018). The immune system responds to the fatty plaques as an abnormal and aberrant event, so it builds a barrier to stop the blood from reaching it and triggers an inflammatory cascade (Libby 2021a). The foam cell mass in vascular walls results in atherosclerotic plaque build-up and helps vessel wall obstruction. This process includes several molecular signaling disruptions such as immune system, inflammation, and oxidative stress as significant molecular switches in the CVDs development (Saigusa, Winkels, and Ley 2020) (Figure 1).

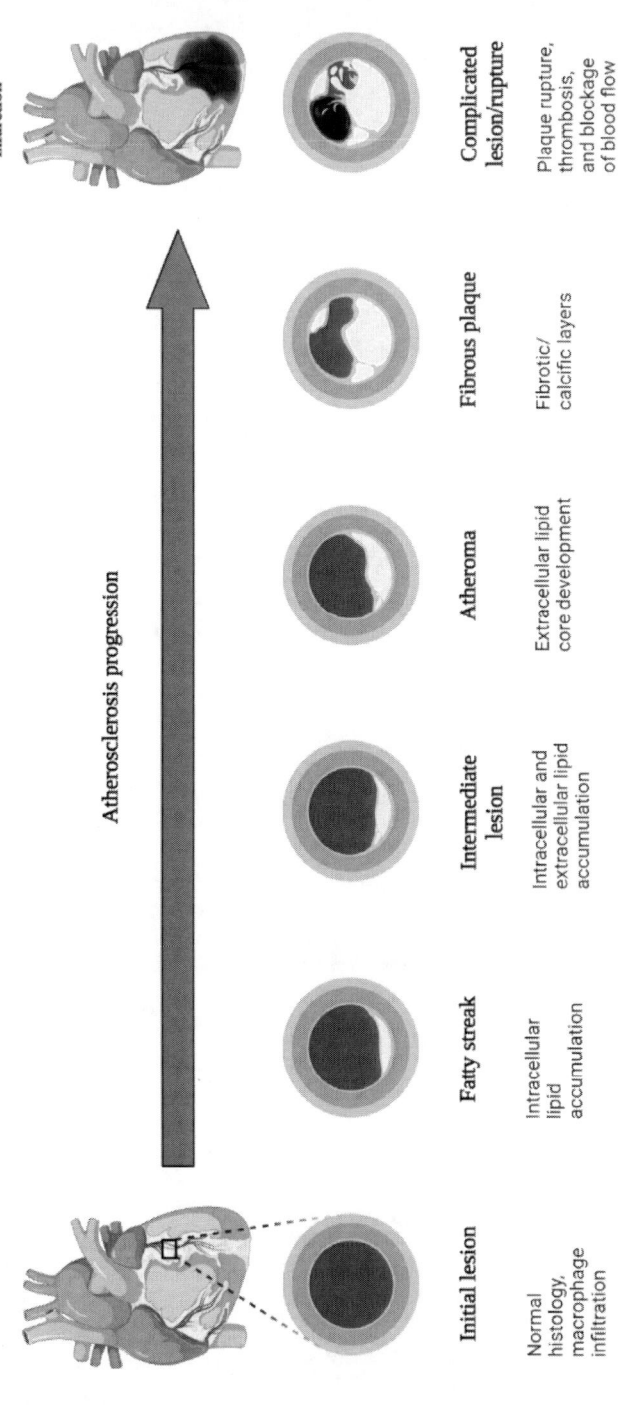

Figure 1. Atherosclerosis progression.

Table 1. Meta-analysis of epidemiological reports of an association between ALA biomarker &/or dietary intake and health outcomes as to cardiovascular diseases

Outcomes analysis	Number of Participates in the Meta-Analysis	ALA source	The average impact of ALA-based HR/RR (95% CIs)	Authors	ref
Total CVD		Dietary Intake & Biomarker	Average RR:0.86,		
Fetal IHD	$n = 251,04$	Dietary Intake & Biomarker	Average RR:0.92,	Pan et al.	(Pan et al., 2012)
Non-fetal IHD		Dietary Intake & Biomarker	Average RR:0.825,		
Stroke		Dietary Intake & Biomarker	Average RR: 0.86,		
CVD mortality	$n = 42,466$	Biomarker	All-cause mortality: HR: 0.99	Harris et al.	(Harris et al., 2021)
Total IHD	$n = 345,202$	Dietary intake	Average RR: 0.91	Wei et al.	(Wei et al., 2018)
Total IHD	$n = 45,637$	Biomarker	Average RR: 1.0	Del Gobbo et al.	(Del Gobbo et al., 2016)

* CVD: cardiovascular diseases, IHD: ischemic heart disease, RR: risk ratio, HR: hazard ratio.

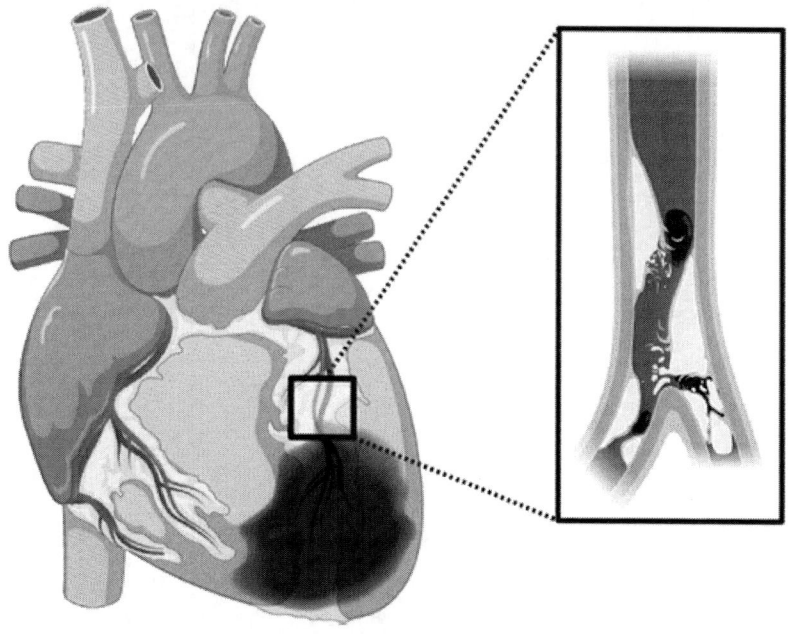

Figure 2. Cardiac arrests caused by atherosclerotic plaque rupture.

However, the plaque may burst if that barrier were to fail. When the substances mix with the blood, a clot forms, and the blood flow is cut off (Figure 2). Most stroke attacks and cardiac arrests are caused by blood atherosclerotic plaque rupture (Libby 2021b).

Based on epidemiological evidence, among behavioral risks, diet plays a significant role in the cardiovascular burden globally (Huang et al., 2022; Hajibabaie et al., 2022). Therefore, managing dietary regimes could be a beneficial strategy to prevent heart arrest and atherosclerotic plaque ruptures (Moriya 2019). Atherosclerosis and plaque formation are linked to dyslipidemia, which causes hyperlipidemia, hypertension, and other high-mortality diseases (Hurtubise et al., 2016). Epidemiological research has shown that changes in lifestyle and food intake may significantly impact lowering dyslipidemia symptoms (Riccardi et al., 2016; Sala-Vila et al., 2022). Regarding lipid metabolism and health, n-3 polyunsaturated fatty acids (PUFA) are one of the essential components of dietary fat (Lenighan, McNulty, and Roche 2019). Eicosapentaenoic acid (EPA) and docosahexaenoic acid (DHA) were shown to have considerably lower Triglyceride (TG) content in a meta-analysis of RCT (Yue et al., 2021). In Table 1, we have shown the overall research about PUFA concentration in blood as a marker, dietary intake as an essential nutrient, and the influence of this bioactive marker on cardiovascular indexes.

Complementary Medicine-Based Alpha-Linolenic Acid

Growing evidence showing the advantages of plant-based diets is focused on a strong interest in learning more about the effects of bioactive substances derived from plant foods and herbal metabolites on cardiometabolic disorders (Hajibabaie et al., 2022; Mullins and Arjmandi 2021; Samtiya et al., 2021; Luo et al., 2021; Szczepańska et al., 2022; Sharif et al., 2020). For example, one of the essential fatty acids in the human diet for the correct function of the physiological body system is α-linolenic acid (ALA) (Yuan et al., 2022). The chemical formula of ALA is 18:3n-3. Moreover, Canonical SMILES of ALA is "CCCCC=CCC=CCCCCCCC(=O) O"; and due to the presence of two double bonds in the structure, classified as PUFAs (Åhlberg 2022) (Figure 3).

Physicochemical parameters of ALA based on the Swiss ADME server indicated that this bioactive compound with a TPSA threshold of 37.30 Å2 could be highly accessible to the gastrointestinal absorption and brain blood-brain barrier permeant. Hence, ALA is described as a drug-likeness component based on Lipinski's pharmacokinetic properties (Daina, Michielin, and Zoete 2017) (Figure 4).

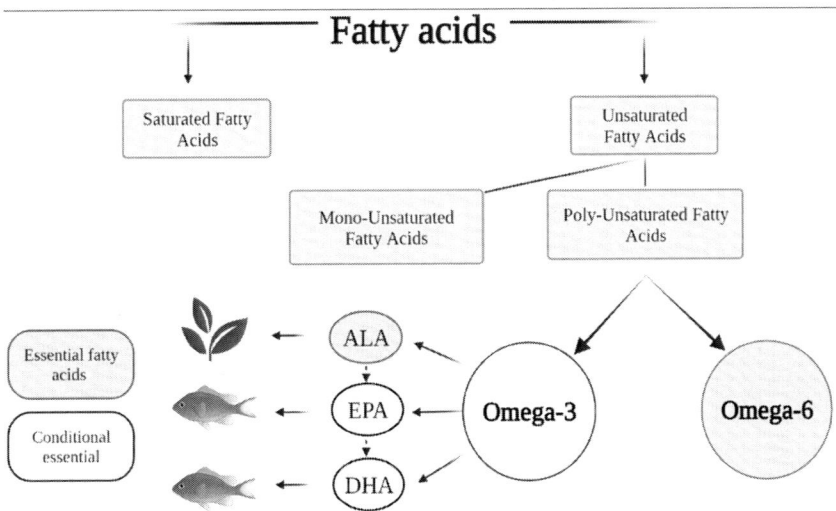

Figure 3. Fatty acids classification is based on saturated & unsaturated fatty acids.

The ALA importance has been overlooked in the past decades due to the focus on the health advantages of dietary EPA and DHA. While EPA and DHA are classified as conditionally essential fatty acids, ALA is an essential unsaturated fatty acid provided by plant sources. On the other hand, ALA might be converted to the EPA in the body, and EPA has the possibility of converting to the DHA (Macaron et al., 2021; Mariamenatu and Abdu 2021; Reimers and Ljung 2019) (Figure 3). ALA derived from plants like flaxseed and walnut is the major source of n-3 PUFA in most geographical areas due to the absence of access to seafood. However, only about 3% of ALA is converted into EPA and DHA in the biological system (Yue et al., 2021).

The function of ALA in lipid metabolism is less well understood than that of EPA and DHA (Punia et al., 2019). The beneficial effects of ALA have been reported in a significant amount of scientific research on many diseases. However, ALA impact on lipid profiles in the blood is controversial (Yue et al., 2021; Haghighatdoost and Hariri 2019). Blood lipid profiles have been identified as major risk factors in epidemiological studies of various diseases (Islam et al., 2019; Abedpoor, Taghian, and Hajibabaie 2022).

ALA has attracted much scientific attention in the last decade (Yang et al., 2021; Hajibabaie et al., 2022). As a result, a systematic attempt has discovered the beneficial biological properties of supplements ALA independent of seafood-derived n-3 PUFAs.

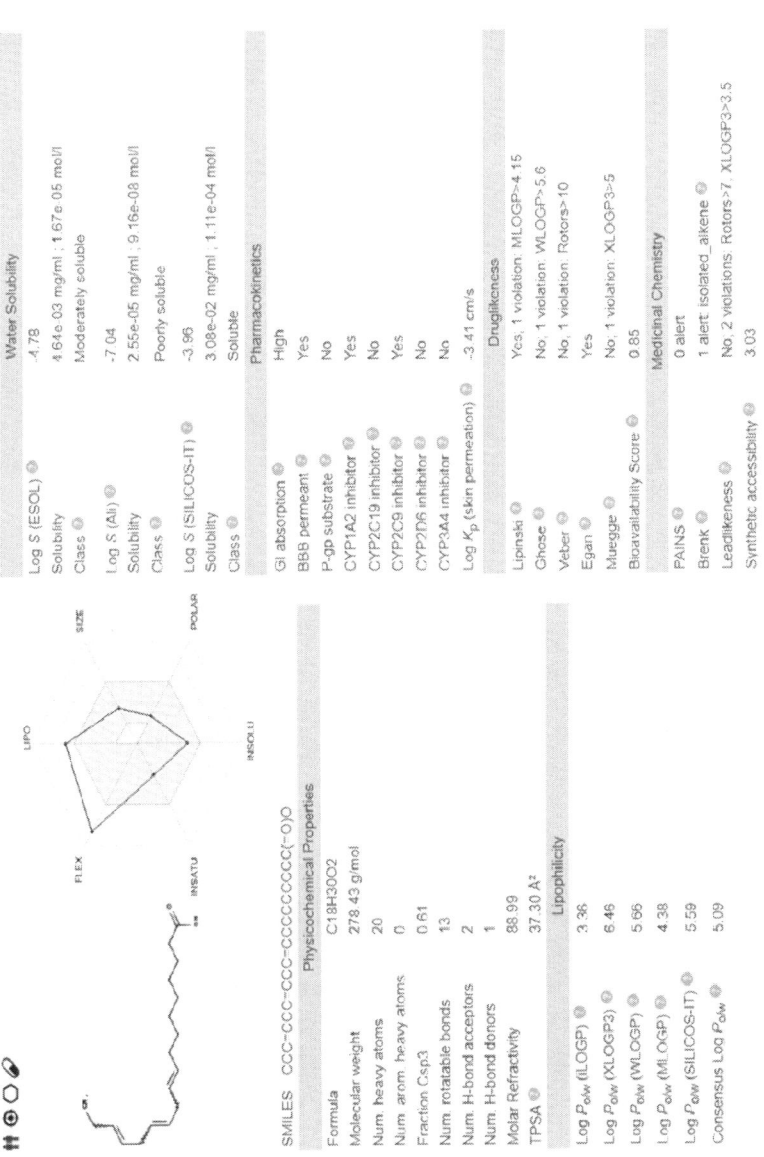

Figure 4. Physicochemical parameters of ALA based on the Swiss ADME server (Daina, Michielin, and Zoete 2017).

The effects of ALA on CVD risk variables have been well-studied and supported by the available data. The effects of ALA on lipids/lipoproteins, blood pressure (BP), inflammatory markers, and other indices of cardiometabolic illness have been the primary focus of the majority of investigations, particularly randomized controlled trials (RCTs). In previous research, supplements and various foods have been used as sources of ALA. Some of these studies compared n-3 PUFAs sourced from marine sources to those found in plants. Based on the evidence, ALA reduces indicators of cardiovascular disease risk. The current body of research is increasing our knowledge of the possible pathways through which ALA reduces the risk of CVD (Abdelhamid et al., 2018; Guasch-Ferré et al., 2018; Ursoniu et al., 2016; Rodriguez-Leyva et al., 2013; Pieters et al., 2019; Hashimoto et al., 2020; Joris et al., 2020).

Supplementation with n-3 PUFA (EPA and DHA) has not been linked to an increased risk of cardiovascular disease in people with diabetes, according to the results of recent randomized controlled trials. In addition, a meta-analysis by Abdelhamid et al. mentioned that ALA intake has no evidence of the impact reduction in blood lipids profiles with six randomized controlled trials (Yue et al., 2021).

Although many people ignore fish and seafood for multiple reasons (Vegetarianism/Veganism, taste, sustainability hypersensitivity, and concern about heavy metals and other contaminants), there is a major demand for natural resources to meet consumption needs. We present a new record from epidemiologic studies, RCTs, and meta-analyses on the association between ALA intake and CVDs, as well as evidence on intermediate risk factors and cardiometabolic risk indicators such as type 2 diabetes (T2D), metabolic syndrome (MetS), and obesity in Table 1.

The ischemic heart disease (IHD) risk and CVD mortality may be attenuated by switching from saturated fatty acids (SFAs) to PUFAs dietary, as recommended in the recent edition of Dietary Guidelines for Americans (2020-2025) (Phillips 2021). In addition, an increase in dietary ALA has been linked to potential health benefits in many recent studies and clinical publications (Committee 2020). Based on the literature review and dietary guidelines, ALA bioactive compounds' safe dose and adequate intake have been determined at 1.1-1.6 g/day (Phillips 2021).

It is worth noting that significant discrepancies exist in the research due to differences in study or prescription supplementary length, dosage, ethnicity, gender, and dietary intake (Sala-Vila et al., 2022). These results are further limited by the fact that trials differed in the form of ALA used (capsules containing oil, seed oils, or whole meals), which affected the supplement's bioavailability. Nevertheless, based on the observational research in a meta-analysis published in 2012, the rather decreased risk of total CVD was related to higher intakes of dietary ALA (RR: 0.90; 95% CI: 0.81, 0.99) (Pan et al., 2012). Moreover, Sala-Vila et al., (Sala-Vila et al., 2016) prospectively investigated the link between supplement ALA and the mortality rate of CVD in participants of the PREvención con DIeta MEDiterránea (PREDIMED) research (n = 7202) to analyze the influence of ALA independently of EPA and DHA (Sala-Vila et al., 2016). According to data, although there was no correlation between dietary ALA and cardiovascular mortality, those who consumed ≥0.7% of their daily calories from ALA had a 28% lower risk of all-cause death, and those who consumed more than 500 mg/day of EPA and DHA had an even lower risk of death (Sala-Vila et al., 2016). The results of this study proposed that the n-3 PUFAs intake from both marine and plant sources may have a synergistic effect in terms of total mortality rate (Pan et al., 2012).

Recently, a systematic meta-analysis using a coordinated methodology employed data records from 17 cohort studies that disclosed LCn-3PUFA isoforms, but not ALA, was strongly linked with significantly low average CVD death rates. This data was published after earlier directional research (de Oliveira Otto et al., 2013; Fretts et al., 2014; Marklund et al., 2015; Harris et al., 2017; Harris et al., 2018).

Computational chemistry methods and pharmacophore modeling revealed that flaxseed-derived natural medicines (SDG, lignans, and ALA) enhance the targeted therapy efficiency of diabetic heart status. New data indicated that flaxseed phytonutrients have the capacity to boost cardiovascular medication by reversing the symptoms of the diabetic heart (Hajibabaie et al., 2022). Therefore, single or combined use of ALA phytonutrient and conventional treatment may be complementary and alternative medicine (Adib-Hajbaghery et al., 2021). Tissue repair may depend, in part, on the inflammatory response to external and endogenous factors such as tissue damage, traumas, and illness. This phenomenon is associated with the production of proinflammatory cytokines, such as interleukin 1 beta (IL-1β), tumor necrosis factor-alpha (TNF-α), and interleukin (IL-6), which are build-up as a result of the accumulation of macrophages (Zhang et al., 2020; Frati et al., 2017).

Conclusion

There is significant literature on the impact of ALA on risk factors in CVDs and CVD's complications. Most studies, predominantly RCTs, focused on lipids/lipoproteins index and their impacts on BP, inflammatory mediators, and other cardio-metabolic complications index. The source of ALA has varied, from supplements to diverse dietary sources. In several of these researches, compare between seafood and plant-based n–3 PUFAs were done. The evidence indicates that ALA has positive impacts on CVD risk indicators. The study to date is improving our knowledge of the possible processes through which ALA might neutralize CVD risk.

References

Abdelhamid, Asmaa S, Tracey J Brown, Julii S Brainard, Priti Biswas, Gabrielle C Thorpe, Helen J Moore, Katherine H O Deane, Fai K AlAbdulghafoor, Carolyn D Summerbell, and Helen V Worthington. 2018. "Omega-3 fatty acids for the primary and secondary prevention of cardiovascular disease." *Cochrane Database of Systematic Reviews* (11).

Abedpoor, Navid, Farzaneh Taghian, and Fatemeh Hajibabaie. 2022. "Physical activity ameliorates the function of organs via adipose tissue in metabolic diseases." *Acta histochemica* 124 (2): 151844.

Adib-Hajbaghery, Mohsen, Mojtaba Fattahi Ardakani, Ahmad Sotoudeh, and Ali Asadian. 2021. "Prevalence of complementary and alternative medicine (CAM) among diabetic patients in Eastern Mediterranean country members of the World Health Organization (WHO): A review." *Journal of Herbal Medicine* 29: 100476.

Åhlberg, Mauri K. 2022. "An update of Åhlberg (2021a): A profound explanation of why eating green (wild) edible plants promote health and longevity." *Food frontiers* 3 (3): 366-379.

Committee, Dietary Guidelines Advisory. 2020. "Scientific report of the 2020 Dietary Guidelines Advisory Committee: advisory report to the Secretary of Agriculture and the Secretary of Health and Human Services." *Agricultural Research Service* 14.

Daina, Antoine, Olivier Michielin, and Vincent Zoete. 2017. "SwissADME: a free web tool to evaluate pharmacokinetics, drug-likeness and medicinal chemistry friendliness of small molecules." *Scientific reports* 7 (1): 42717.

de Oliveira Otto, Marcia C, Jason HY Wu, Ana Baylin, Dhananjay Vaidya, Stephen S Rich, Michael Y Tsai, David R Jacobs Jr, and Dariush Mozaffarian. 2013. "Circulating and dietary omega-3 and omega-6 polyunsaturated fatty acids and incidence of CVD in the Multi-Ethnic Study of Atherosclerosis." *Journal of the American Heart Association* 2 (6): e000506.

Del Gobbo, Liana C, Fumiaki Imamura, Stella Aslibekyan, Matti Marklund, Jyrki K Virtanen, Maria Wennberg, Mohammad Y Yakoob, Stephanie E Chiuve, Luicito Dela Cruz, and Alexis C Frazier-Wood. 2016. "ω-3 polyunsaturated fatty acid biomarkers and coronary heart disease: pooling project of 19 cohort studies." *JAMA internal medicine* 176 (8): 1155-1166.

Frati, Giacomo, Leonardo Schirone, Isotta Chimenti, Derek Yee, Giuseppe Biondi-Zoccai, Massimo Volpe, and Sebastiano Sciarretta. 2017. "An overview of the inflammatory signalling mechanisms in the myocardium underlying the development of diabetic cardiomyopathy." *Cardiovascular research* 113 (4): 378-388.

Fretts, Amanda M, Dariush Mozaffarian, David S Siscovick, Colleen Sitlani, Bruce M Psaty, Eric B Rimm, Xiaoling Song, Barbara McKnight, Donna Spiegelman, and Irena B King. 2014. "Plasma phospholipid and dietary α-linolenic acid, mortality, CHD and stroke: the Cardiovascular Health Study." *British journal of nutrition* 112 (7): 1206-1213.

Gaidai, Oleg, Yu Cao, and Stas Loginov. 2023. "Global cardiovascular diseases death rate prediction." *Current Problems in Cardiology*: 101622.

Guasch-Ferré, Marta, Jun Li, Frank B Hu, Jordi Salas-Salvadó, and Deirdre K Tobias. 2018. "Effects of walnut consumption on blood lipids and other cardiovascular risk factors: an updated meta-analysis and systematic review of controlled trials." *The American journal of clinical nutrition* 108 (1): 174-187.

Haghighatdoost, Fahimeh, and Mitra Hariri. 2019. "Does alpha-lipoic acid affect lipid profile? A meta-analysis and systematic review on randomized controlled trials." *European journal of pharmacology* 847: 1-10.

Hajibabaie, Fatemeh, Navid Abedpoor, Kamran Safavi, and Farzaneh Taghian. 2022. "Natural remedies medicine derived from flaxseed (secoisolariciresinol diglucoside, lignans, and α-linolenic acid) improve network targeting efficiency of diabetic heart conditions based on computational chemistry techniques and pharmacophore modeling." *Journal of Food Biochemistry*: e14480.

Hajibabaie, Fatemeh, Shirin Kouhpayeh, Mina Mirian, Ilnaz Rahimmanesh, Maryam Boshtam, Ladan Sadeghian, Azam Gheibi, Hossein Khanahmad, and Laleh Shariati. 2020. "MicroRNAs as the actors in the atherosclerosis scenario." *Journal of physiology and biochemistry* 76: 1-12.

Harris, William S, Juhua Luo, James V Pottala, Mark A Espeland, Karen L Margolis, Joann E Manson, Lu Wang, Theodore M Brasky, and Jennifer G Robinson. 2017. "Red blood cell polyunsaturated fatty acids and mortality in the Women's Health Initiative Memory Study." *Journal of clinical lipidology* 11 (1): 250-259. e5.

Harris, William S, Nathan L Tintle, Mark R Etherton, and Ramachandran S Vasan. 2018. "Erythrocyte long-chain omega-3 fatty acid levels are inversely associated with mortality and with incident cardiovascular disease: The Framingham Heart Study." *Journal of clinical lipidology* 12 (3): 718-727. e6.

Harris, William S, Nathan L Tintle, Fumiaki Imamura, Frank Qian, Andres V Ardisson Korat, Matti Marklund, Luc Djoussé, Julie K Bassett, Pierre-Hugues Carmichael, and Yun-Yu Chen. 2021. "Blood n-3 fatty acid levels and total and cause-specific mortality from 17 prospective studies." *Nature communications* 12 (1): 2329.

Hashimoto, Michio, Yoko Tanabe, Shahdat Hossain, Kentaro Matsuzaki, Miho Ohno, Setsushi Kato, Masanori Katakura, and Osamu Shido. 2020. "Intake of alpha-linolenic acid-rich Perilla frutescens leaf powder decreases home blood pressure and serum oxidized low-density lipoprotein in Japanese adults." *Molecules* 25 (9): 2099.

Hoseini, Zahra, Fatemeh Sepahvand, Bahman Rashidi, Amirhossein Sahebkar, Aria Masoudifar, and Hamed Mirzaei. 2018. "NLRP3 inflammasome: Its regulation and involvement in atherosclerosis." *Journal of cellular physiology* 233 (3): 2116-2132.

Huang, Wen-Ching, Chun-Liang Tung, Yu-Chen SH Yang, I-Hsuan Lin, Xin Er Ng, and Yu-Tang Tung. 2022. "Endurance exercise ameliorates Western diet–induced atherosclerosis through modulation of microbiota and its metabolites." *Scientific Reports* 12 (1): 3612.

Hurtubise, Jessica, Krystie McLellan, Kevin Durr, Oluwadara Onasanya, Daniel Nwabuko, and Joseph Fomusi Ndisang. 2016. "The different facets of dyslipidemia and hypertension in atherosclerosis." *Current atherosclerosis reports* 18: 1-12.

Islam, Md Ashraful, Mohammad Nurul Amin, Shafayet Ahmed Siddiqui, Md Parvez Hossain, Farhana Sultana, and Md Ruhul Kabir. 2019. "Trans fatty acids and lipid profile: A serious risk factor to cardiovascular disease, cancer and diabetes." *Diabetes & Metabolic Syndrome: Clinical Research & Reviews* 13 (2): 1643-1647.

Joris, Peter J, Richard Draijer, Dagmar Fuchs, and Ronald P Mensink. 2020. "Effect of α-linolenic acid on vascular function and metabolic risk markers during the fasting and postprandial phase: a randomized placebo-controlled trial in untreated (pre-) hypertensive individuals." *Clinical Nutrition* 39 (8): 2413-2419.

Laslett, Lawrence J, Peter Alagona, Bernard A Clark, Joseph P Drozda, Frances Saldivar, Sean R Wilson, Chris Poe, and Menolly Hart. 2012. "The worldwide environment of cardiovascular disease: prevalence, diagnosis, therapy, and policy issues: a report from the American College of Cardiology." *Journal of the American College of Cardiology* 60 (25S): S1-S49.

Lenighan, Yvonne M, Breige A McNulty, and Helen M Roche. 2019. "Dietary fat composition: replacement of saturated fatty acids with PUFA as a public health strategy, with an emphasis on α-linolenic acid." *Proceedings of the Nutrition Society* 78 (2): 234-245.

Libby, Peter. 2021a. "The changing landscape of atherosclerosis." *Nature* 592 (7855): 524-533.

Libby, Peter. 2021b. "The changing Nature of atherosclerosis: what we thought we knew, what we think we know, and what we have to learn." *European Heart Journal* 42 (47): 4781-4782.

Luo, Jing, Hongwei Si, Zhenquan Jia, and Dongmin Liu. 2021. "Dietary anti-aging polyphenols and potential mechanisms." *Antioxidants* 10 (2): 283.

Macaron, Tony, Kelly Virecoulon Giudici, Gene L Bowman, Alan Sinclair, Elie Stephan, Bruno Vellas, and Philipe de Souto Barreto. 2021. "Associations of Omega-3 fatty acids with brain morphology and volume in cognitively healthy older adults: A narrative review." *Ageing research reviews* 67: 101300.

Mariamenatu, Abeba Haile, and Emebet Mohammed Abdu. 2021. "Overconsumption of omega-6 polyunsaturated fatty acids (PUFAs) versus deficiency of omega-3 PUFAs in modern-day diets: the disturbing factor for their "balanced antagonistic metabolic functions" in the human body." *Journal of Lipids* 2021: 1-15.

Marklund, Matti, Karin Leander, Max Vikström, Federica Laguzzi, Bruna Gigante, Per Sjögren, Tommy Cederholm, Ulf de Faire, Mai-Lis Hellenius, and Ulf Riserus. 2015. "Polyunsaturated fat intake estimated by circulating biomarkers and risk of cardiovascular disease and all-cause mortality in a population-based cohort of 60-year-old men and women." *Circulation* 132 (7): 586-594.

Moriya, Junji. 2019. "Critical roles of inflammation in atherosclerosis." *Journal of cardiology* 73 (1): 22-27.

Mullins, Amy P, and Bahram H Arjmandi. 2021. "Health benefits of plant-based nutrition: focus on beans in cardiometabolic diseases." *Nutrients* 13 (2): 519.

Pan, An, Mu Chen, Rajiv Chowdhury, Jason HY Wu, Qi Sun, Hannia Campos, Dariush Mozaffarian, and Frank B Hu. 2012. "α-Linolenic acid and risk of cardiovascular disease: a systematic review and meta-analysis." *The American journal of clinical nutrition* 96 (6): 1262-1273.

Phillips, Jennan A. 2021. "Dietary guidelines for Americans, 2020–2025." *Workplace health & safety* 69 (8): 395-395.

Pieters, Dorien J, Peter L Zock, Dagmar Fuchs, and Ronald P Mensink. 2019. "Effect of α-linolenic acid on 24-h ambulatory blood pressure in untreated high-normal and stage I hypertensive subjects." *British Journal of Nutrition* 121 (2): 155-163.

Punia, Sneh, Kawaljit Singh Sandhu, Anil Kumar Siroha, and Sanju Bala Dhull. 2019. "Omega 3-metabolism, absorption, bioavailability and health benefits–A review." *PharmaNutrition* 10: 100162.

Reimers, Arne, and Hanna Ljung. 2019. "The emerging role of omega-3 fatty acids as a therapeutic option in neuropsychiatric disorders." *Therapeutic advances in psychopharmacology* 9: 2045125319858901.

Riccardi, Gabriele, Olga Vaccaro, Giuseppina Costabile, and Angela A Rivellese. 2016. "How well can we control dyslipidemias through lifestyle modifications?" *Current cardiology reports* 18: 1-9.

Rodriguez-Leyva, Delfin, Wendy Weighell, Andrea L Edel, Renee LaVallee, Elena Dibrov, Reinhold Pinneker, Thane G Maddaford, Bram Ramjiawan, Michel Aliani, and Randolph Guzman. 2013. "Potent antihypertensive action of dietary flaxseed in hypertensive patients." *Hypertension* 62 (6): 1081-1089.

Saigusa, Ryosuke, Holger Winkels, and Klaus Ley. 2020. "T cell subsets and functions in atherosclerosis." *Nature Reviews Cardiology* 17 (7): 387-401.

Sala-Vila, Aleix, Jennifer Fleming, Penny Kris-Etherton, and Emilio Ros. 2022. "Impact of α-linolenic acid, the vegetable ω-3 fatty acid, on cardiovascular disease and cognition." *Advances in Nutrition* 13 (5): 1584-1602.

Sala-Vila, Aleix, Marta Guasch-Ferré, Frank B Hu, Ana Sánchez-Tainta, Monica Bulló, Mercè Serra-Mir, Carmen López-Sabater, Jose V Sorlí, Fernando Arós, and Miquel Fiol. 2016. "Dietary α-linolenic acid, marine ω-3 fatty acids, and mortality in a population with high fish consumption: findings from the Prevención Con Dieta Mediterránea (PREDIMED) Study." *Journal of the American Heart Association* 5 (1): e002543.

Samtiya, Mrinal, Rotimi E Aluko, Tejpal Dhewa, and José Manuel Moreno-Rojas. 2021. "Potential health benefits of plant food-derived bioactive components: An overview." *Foods* 10 (4): 839.

Sharif, Hina, Muhammad Sajid Hamid Akash, Kanwal Rehman, Kanwal Irshad, and Imran Imran. 2020. "Pathophysiology of atherosclerosis: Association of risk factors and treatment strategies using plant-based bioactive compounds." *Journal of Food Biochemistry* 44 (11): e13449.

Szczepańska, Elżbieta, Agnieszka Białek-Dratwa, Barbara Janota, and Oskar Kowalski. 2022. "Dietary Therapy in Prevention of Cardiovascular Disease (CVD)—Tradition or Modernity? A Review of the Latest Approaches to Nutrition in CVD." *Nutrients* 14 (13): 2649.

Tsao, Connie W, Aaron W Aday, Zaid I Almarzooq, Alvaro Alonso, Andrea Z Beaton, Marcio S Bittencourt, Amelia K Boehme, Alfred E Buxton, April P Carson, and Yvonne Commodore-Mensah. 2022a. "Heart disease and stroke statistics—2022 update: a report from the American Heart Association." *Circulation* 145 (8): e153-e639.

Tsao, C W, A W Aday, Z I Almarzooq, A Alonso, A Z Beaton, M S Bittencourt, A K Boehme, A E Buxton, A P Carson, and Y Commodore-Mensah. 2022b. "American heart association council on epidemiology and prevention statistics committee and stroke statistics subcommittee." *Heart disease and stroke statistics-2022 update: A report from the American Heart Association. Circulation* 145 (8): e153-e639.

Ursoniu, Sorin, Amirhossein Sahebkar, Florina Andrica, Corina Serban, Maciej Banach, Lipid, and Blood Pressure Meta-analysis Collaboration. 2016. "Effects of flaxseed supplements on blood pressure: A systematic review and meta-analysis of controlled clinical trial." *Clinical nutrition* 35 (3): 615-625.

Wei, Jingkai, Ruixue Hou, Yuzhi Xi, Alysse Kowalski, Tiansheng Wang, Zhi Yu, Yirui Hu, Eeshwar K Chandrasekar, Hao Sun, and Mohammed K Ali. 2018. "The association and dose–response relationship between dietary intake of α-linolenic acid and risk of CHD: a systematic review and meta-analysis of cohort studies." *British Journal of Nutrition* 119 (1): 83-89.

Yang, Jing, Chaoting Wen, Yuqing Duan, Qianchun Deng, Dengfeng Peng, Haihui Zhang, and Haile Ma. 2021. "The composition, extraction, analysis, bioactivities, bioavailability and applications in food system of flaxseed (Linum usitatissimum L.) oil: A review." *Trends in Food Science & Technology* 118: 252-260.

Yuan, Qianghua, Fan Xie, Wei Huang, Mei Hu, Qilu Yan, Zemou Chen, Yan Zheng, and Li Liu. 2022. "The review of alpha-linolenic acid: Sources, metabolism, and pharmacology." *Phytotherapy Research* 36 (1): 164-188.

Yue, Hao, Bin Qiu, Min Jia, Wei Liu, Xiao-fei Guo, Na Li, Zhi-xiang Xu, Fang-ling Du, Tongcheng Xu, and Duo Li. 2021. "Effects of α-linolenic acid intake on blood lipid profiles : a systematic review and meta-analysis of randomized controlled trials." *Critical Reviews in Food Science and Nutrition* 61 (17): 2894-2910.

Zhang, Xin-Yue, Zheng Huang, Qing-Jie Li, Guo-Qiang Zhong, Jian-Jun Meng, Dong-Xiao Wang, and Rong-Hui Tu. 2020. "Ischemic postconditioning attenuates the inflammatory response in ischemia/reperfusion myocardium by upregulating miR-499 and inhibiting TLR2 activation." *Molecular Medicine Reports* 22 (1): 209-218.

Biographical Sketch

Name: *Fatemeh Hajibabaie*

Affiliation: Department of Biology, Faculty of Basic Sciences, Shahrekord Branch, Islamic Azad University, Shahrekord, Iran.
Department of Physiology, Medicinal Plants Research Center, Isfahan (Khorasgan) Branch, Islamic Azad University, Isfahan, Iran.

Education: Molecular genetics.

Business Address: Isfahan (Khorasgan) Branch, Islamic Azad University, Isfahan, Iran.

Research and Professional Experience: Genetics, System biology, Bioinformatics, Drug Design, Drug Delivery, Non-Coding RNAs, Cancer.

Professional Appointments: Researcher.

Honors: Best Poster in Royan International twin congress, Reproductive Biomedicine & Stem Cell. Protective approaches of Fraxinus excelsior compounds on the Implantation based infertility via bioinformatics and chemoinformatic analysis.

Publications from the Last 3 Years:

1. Fatemeh Hajibabaie, Shirin Kouhpayeh, Mina Mirian, Ilnaz Rahimmanesh, Maryam Boshtam, Ladan Sadeghian, Azam Gheibi, Hossein Khanahmad and Laleh Shariati. (2019). MicroRNAs as the actors in the atherosclerosis scenario. *Journal of Physiology and Biochemistry.*
2. Navid Abedpoor, Farzaneh Taghian, Fatemeh Hajibabaie. (2022). Physical activity ameliorates the function of organs via adipose tissue in metabolic diseases. *Acta histochemical.*
3. Fatemeh Hajibabaie, Navid Abedpoor, Nazanin Asareh, Mohammad Amin Tabatabaiefar, Ali Zarrabi and Laleh Shariati. (2022). A cocktail of microRNAs as an advance diagnostic signature in stomach-colorectal cancers hallmarks incidence: a systematic review. *Personal Medicine.*
4. Navid Abedpoor, Farzaneh Taghian, Fatemeh Hajibabaie. (2022). Cross Brain-Gut Analysis Highlighted Hub Genes and LncRNAs Networks Differentially Modified

During Leucine Consumption and Endurance Exercise in Mice with Depression Like Behaviors. *Molecular Neurobiology*.

5. Fatemeh Hajibabaie, Farzaneh Taghian, Navid Abedpoor, Kamran Safavi. (2023). A Cocktail of Polyherbal Bioactive Compounds and Regular Mobility Training as Senolytic Approaches in Age-dependent Alzheimer's: the In Silico Analysis, Lifestyle Intervention in Old Age. *Journal of Molecular Neuroscience*.

6. Fatemeh Hajibabaie, Navid Abedpoor, Kamran Safavi, Farzaneh Taghian. (2022). Natural remedies medicine derived from flaxseed (Secoisolariciresinol diglucoside, lignans, and α-linolenic acid) improve network targeting efficiency of diabetic heart conditions based on computational chemistry techniques and pharmacophore modeling. *Journal of Food Biochemistry*.

7. Fatemeh Hajibabaie, Faranak Aali, Navid Abedpoor. (2022). Pathomechanisms of non-coding RNAs and hub genes related to the oxidative stress in diabetic complications. *F1000Research*.

8. Fatemeh Hajibabaie, Navid Abedpoor. (2023). *The Importance of Hub Genes and Genetically and Epigenetically Modulators as Potent Biomarkers in the Prognosis, Diagnosis, and Therapeutic Monitoring of Colorectal Cancer. Horizons in Cancer Research*. Volume 85. Nova Science Publishers

Index

α

α-Linolenic, vii, 1, 3, 4, 5, 6, 10, 12, 13, 14, 15, 17, 19, 36, 37, 49, 57, 58, 60, 70, 72, 76, 83, 89, 90, 91, 92, 94

A

acid(s), vii, 1, 3, 4, 5, 6, 7, 10, 11, 12, 13, 14, 15, 17, 19, 21, 23, 24, 25, 28, 29, 30, 31, 32, 33, 34, 35, 36, 37, 38, 39, 40, 41, 45, 46, 47, 48, 49, 50, 51, 52, 53, 54, 55, 56, 57, 58, 60, 62, 63, 64, 65, 69, 70, 71, 72, 73, 74, 76, 77, 82, 83, 84, 86, 88, 89, 90, 91, 92, 94
adaptive, 22, 23, 26, 29, 30, 33, 34, 35
alpha, 11, 14, 21, 22, 46, 48, 49, 50, 54, 56, 57, 58, 62, 65, 69, 70, 77, 78, 83, 87, 89, 90, 92
alpha-linolenic acid, 21, 22, 46, 48, 49, 50, 54, 58, 62, 65, 78, 90
atherosclerosis, 14, 35, 59, 77, 78, 79, 80, 82, 88, 89, 90, 91, 92, 93

C

cardiovascular, vii, 12, 61, 65, 77, 78, 79, 81, 82, 86, 87, 88, 89, 90, 91, 92
cardiovascular diseases (CVDs), 61, 77, 78, 79, 81, 86, 88, 89
cell(s), 6, 10, 13, 14, 16, 18, 21, 22, 23, 24, 25, 26, 27, 28, 29, 30, 31, 32, 33, 34, 35, 36, 37, 38, 39, 40, 41, 46, 47, 48, 49, 50, 51, 52, 53, 54, 55, 56, 57, 58, 59, 63, 64, 67, 68, 69, 70, 71, 72, 73, 74, 75, 78, 79, 89, 91, 93

colorectal cancer, vii, 16, 18, 19, 42, 57, 60, 61, 62, 63, 64, 68, 69, 70, 71, 72, 73, 74, 75, 76, 93, 94
complementary, 5, 54, 56, 57, 83, 87, 88
complication(s), 5, 14, 16, 17, 18, 19, 42, 60, 75, 76, 77, 78, 88, 94
coronary heart disease, 77, 89
cytokines and chemokines, 22, 28, 34

D

deficiency, 3, 13, 77, 90
dendritic, 22, 30, 36, 37, 40, 41
diabetes, 4, 10, 12, 14, 16, 17, 18, 19, 42, 43, 61, 65, 69, 75, 76, 78, 86, 90

E

endoplasmic reticulum stress, 1, 2, 11, 12, 13, 64, 69
epithelium, 25, 26, 35

F

fatty, 1, 2, 3, 4, 5, 7, 11, 12, 13, 14, 15, 21, 24, 25, 28, 29, 30, 31, 32, 33, 34, 35, 36, 37, 38, 39, 40, 41, 45, 46, 47, 48, 49, 50, 51, 54, 55, 56, 57, 58, 62, 63, 64, 65, 69, 70, 71, 72, 77, 79, 82, 83, 84, 86, 88, 89, 90, 91

H

hallmarks, 16, 19, 42, 45, 60, 76, 77, 93
hepatic, vii, 1, 2, 4, 5, 7, 9, 11, 12, 13, 14, 15
hepatic lipid, 2, 13

Index

hepatic steatosis, vii, 1, 4, 5, 7, 9, 11, 12, 15

I

immune, 4, 21, 22, 23, 25, 26, 28, 29, 30, 31, 33, 34, 35, 36, 38, 40, 47, 55, 63, 64, 72, 79
immune response, 21, 22, 25, 28, 30, 33, 35, 38, 40, 63
immune system, 4, 21, 22, 29, 30, 33, 40, 47, 55, 64, 79
improving, vii, 1, 5, 9, 64, 88
inflammation, 1, 3, 4, 7, 11, 13, 15, 17, 18, 19, 23, 27, 28, 35, 36, 37, 39, 40, 41, 43, 45, 46, 47, 49, 62, 63, 69, 71, 73, 74, 75, 76, 78, 79, 91
innate and adaptive immune response, vii, 22
insulin, 1, 4, 5, 10, 12, 13, 14, 15, 39, 52, 68

L

lifestyle(s), 4, 5, 15, 17, 18, 19, 41, 45, 46, 48, 56, 60, 61, 75, 76, 78, 82, 91, 94
linolenic acid, vii, 2, 5, 7, 10, 14, 21, 22, 45, 49, 50, 51, 53, 55, 56, 57, 58, 61, 62, 65, 70, 72, 77, 83, 89, 91, 92
lipid, 2, 5, 6, 10, 13, 15, 22, 23, 35, 36, 37, 38, 39, 40, 57, 62, 65, 69, 71, 77, 78, 79, 82, 84, 89, 90, 92
liver, 1, 2, 3, 4, 5, 7, 9, 10, 11, 12, 13, 14, 15, 69
liver disease(s), 2, 3, 4, 5, 7, 9, 10, 11, 12, 13, 14, 15

M

macrophages, 22, 27, 28, 31, 32, 35, 36, 39, 40, 67, 69, 72, 74, 87
maintenance, 45, 46, 49
medication(s), 5, 45, 47, 87
medicine, 12, 16, 17, 18, 19, 35, 36, 41, 42, 54, 55, 56, 57, 58, 60, 70, 71, 74, 75, 76, 78, 83, 87, 88, 89, 92, 93, 94

modern society, 61
modulate, vii, 6, 24, 28, 40, 61, 62, 65
molecular physiologists, 78

N

natural, 7, 12, 17, 19, 22, 33, 45, 47, 48, 50, 52, 54, 56, 57, 60, 70, 76, 77, 86, 87, 89, 94
neurodegeneration, 45, 48
neurogenesis, vii, 45, 46, 47, 48, 49, 50, 51, 52, 54, 55, 56, 57, 58
neutrophil, 29, 35, 40
nonalcoholic, 2, 4, 5, 11, 12, 13, 14, 15

O

offset, 77
omega, 1, 12, 13, 21, 24, 25, 28, 29, 30, 32, 33, 34, 35, 36, 37, 38, 39, 40, 49, 50, 54, 55, 56, 58, 88, 89, 90, 91

P

pathomechanism, vii, 61, 62
preventing, vii, 1, 2, 5, 33, 64

R

remedies, 12, 17, 19, 56, 60, 70, 76, 77, 89, 94
response(s), 5, 6, 11, 13, 14, 22, 24, 26, 27, 28, 30, 33, 35, 36, 46, 56, 63, 67, 72, 87, 92
role, 3, 4, 5, 7, 10, 14, 26, 28, 29, 30, 31, 33, 34, 35, 46, 47, 49, 50, 51, 52, 54, 55, 56, 58, 62, 64, 70, 72, 74, 82, 91

S

statuses, 45, 78
steatosis, 1, 4, 5

T

type 2 diabetes, 1, 2, 4, 10, 14, 86